# Value and Risk Management

# Value and Risk Management

## A guide to best practice

Michael Dallas, MA (Cantab), MICE, FIVM

**Blackwell**
Publishing

CIOB

© 2006 Michael Dallas

Editorial offices:
Blackwell Publishing Ltd, 9600 Garsington Road, Oxford OX4 2DQ, UK
  Tel: +44 (0) 1865 776868
Blackwell Publishing Inc., 350 Main Street, Malden, MA 02148-5020, USA
  Tel: +1 781 388 8250
Blackwell Science Asia Pty Ltd, 550 Swanston Street, Carlton, Victoria 3053, Australia
  Tel: +61 (0)3 8359 1011

First published 2006 by Blackwell Publishing Ltd

4    2009

Library of Congress Cataloging-in-Publication Data

Dallas, Michael.
Value and risk management: a guide to best practice/by Michael Dallas.
    p. cm.
  Includes bibliographical references and index.
  ISBN: 978-14051-2069-2 (alk. paper)
1. Construction industry–Risk management. 2. Value analysis (Cost control)
3. Project management. I. Title.

HD9715.A2D314 2005
658.4'04–dc22

                                        2005010287


A catalogue record for this title is available from the British Library

Set in 10/13 pt Grotesque MT
by Newgen Imaging Systems (P) Ltd, Chennai, India
Printed in Singapore by COS Printers Pte Ltd

For further information on Blackwell Publishing, visit our website:
www.blackwellpublishing.com

# Contents

# Foreword

This book raises the subjects of value and risk management above the levels of simple cost reduction and their application primarily to construction activities with which they are often, rather narrowly, associated. At the same time, it improves accessibility to what are often considered to be fairly complicated processes.

Construction projects are about more than simply delivering buildings. They must reflect the long-term business needs of those who commission them and deliver the expected benefits.

In this context, value lies in the effectiveness with which the benefits are delivered. Effective delivery requires that the industry clearly understands the long-term needs of the clients, and delivers them efficiently and economically. The integrated project team optimises the efficiency of delivery through the supply chain. Value management complements efficient delivery by ensuring that efforts are deployed towards delivering the right buildings. It helps create affordable buildings that are productive and pleasant places in which to work, that are easy to use and maintain, and contribute positively to the environment and communities in which they are situated.

In contrast, effective risk management ensures that value is not eroded by avoidable mishaps or uncertainties. To achieve this aim, the team must from the outset ensure that the conditions are put in place to enable successful project delivery. This means looking wider than the construction phase at the way in which the project is set-up and procured. It means actively managing the risks from inception to use.

In this book, Michael Dallas describes how value management provides the means to articulate and deliver best value, while risk management provides assurance that the set-up and delivery of the project will avoid the destruction of value. They should both be integrated with other project management processes to maximise the likelihood of a successful outcome.

I strongly recommend this book to all those who wish to raise their understanding of value and risk management. It is essential reading for those who want to add value and reduce uncertainty in their development and construction projects.

**Peter Rogers**
Chairman, Strategic Forum for Construction
Chairman, Constructing Excellence
Director, Stanhope plc

# Preface

## Aims

This book aims to change the established mindset in the construction industry from the delivery of a building to the delivery of the expected benefits to those commissioning the building. In order to do this, it describes how value management enables the team to articulate, communicate and maximise value in terms of the benefits, while risk management minimises the uncertainty in delivering them. It goes on to describe the processes and techniques that make this possible.

It is intended to stimulate understanding and interest in the subjects of value and risk management and provide a companion volume to the Chartered Institute of Buildings Code of Practice for Project Management for Construction and Development.

Senior and middle management will benefit most from reading Chapters 1–4. These describe how the use of value and risk management will help them make informed decisions and provide informed briefings to those to whom they entrust the delivery of their projects.

Aspiring and active practitioners will benefit from the many tips and examples of practical experience, which form the main substance of the book.

Students will gain useful insights into the concepts, processes and methods by using the book as a reference manual, thus complementing the academic foundation that they will receive during their university or post-graduate studies.

Many designers find the formality of value and risk management disconcerting. They worry that the processes will undermine the finer qualities of their designs. This book should allay those fears and help them embrace the positive contributions the processes have to offer.

# Filling a gap

Most books on value and risk management are long on technique but short on sound, practical advice. This book provides both, but is not an academic treatise. While it draws on the themes expressed in many publications, it is based on the author's considerable practical experience. It describes what works and what does not in the United Kingdom, as well as elsewhere in the world.

The author believes that value and risk are inextricably linked. Both must be explicitly addressed in a coordinated programme for project success. All projects involve people whose diverse interests must be addressed and reconciled. No project happens in isolation. The project team must consider their activities in relation to the environment in which they are operating. The approaches to value and risk management described in these pages address all these interrelated issues and provides practical and tested advice on how to manage the complex interactions of value, risk, people and the project environment.

Therefore the book fills a gap left by other publications on these subjects.

# Structure

The book takes the reader on a journey from first principles to the applications of value and risk management and builds an understanding on how to apply the processes, methods and techniques. It is structured to provide increasing levels of detail from beginning to end.

Chapter 1   Explains why value and risk management are necessary and what they contribute to achieving successful projects.

Chapter 2   Introduces the background and basic principles of value management.

Chapter 3   Describes the background and basic principles of risk management.

Chapter 4   Explains how the two disciplines naturally integrate into a single programme of activities, throughout the life of a project and, in doing so, complement each other.

| Chapter 5 | Describes the vital roles that people play in the processes. |
| Chapter 6 | Explains the underlying concepts of value and risk and introduces British Standards that are current. |
| Chapter 7 | Contains guest contributions to provide a glimpse of best practice from practitioners in other sectors of the economy. |
| Chapter 8 | Explains how the different types of study may be applied at key project stages. |
| Chapters 9–11 | Describes commonly used techniques. |
| Chapter 12 | Contains examples of forms, checklists and other tools to help deliver the processes effectively |

## Interrelationship of value and risk

Value management is about clearly articulating what represents value in terms of the project benefits and then linking these to the most cost-effective design solutions. Risk management is about identifying causes of uncertainty and what can go wrong, and then putting in place activities to minimise the adverse impact on the project. Both activities complement each other. Value management can reduce risk. Risk management can provide opportunities to increase value. It can also avoid the destruction of value.

*Unless value is clearly articulated at the outset and, later, delivered in the finished product, it will not be maximised.*

*Unless risk is identified and its consequences controlled, value will be destroyed.*

## Leadership and structured methodology

Both disciplines contain specific techniques designed to achieve the desired outcomes. They are not haphazard. The techniques are best applied by those who are familiar with their use. The author uses the term 'study leader' to describe such individuals. Both disciplines rely on the collaborative efforts of the people engaged in delivering the project, the project team. The book describes how the study leader can assist the team to work together constructively.

## Concepts explained

The book describes the underlying concepts of value and risk management and how they relate to each other. It describes the different issues that must be addressed at all stages throughout the life of a project. It suggests practical ways in which to handle people and organisations with different interests. It describes how to build a culture that understands and manages value and risk both through formal studies and intuitive advice. It explains how to introduce value and risk management into an organisation to improve its competitiveness and productivity. It reviews commonly used and effective techniques and suggests how these may be adapted to suit individuals' styles and circumstances.

## Case studies

Throughout, the author draws on his library of success (and some failure) stories to bring the messages to life by means of case studies and anecdotes.

## Best practice

Best practice is not static. Experience from other industries and cultures provide a rich source of continuous improvement. The book explores value and risk management practice in sectors outside construction and suggests how these may influence future best practice in construction.

## Techniques

The book describes the use of clearly signposted techniques and checklists to support the study leader in conducting effective value and risk studies. This level of detail is intended to provide a code of best practice to students of the subjects and practitioners alike.

# About the Author

Michael Dallas grew up on a farm in Sussex in the period of post war austerity in the 1950s. He was educated at Sherborne School and Cambridge University, gaining degrees in Natural Sciences and Chemical Engineering. This background gave him a very practical view on life, appreciating value, abhorring waste, needing to innovate and manage day to day risk.

His career has involved building Dams, Pipelines and Water Treatment Works, manufacturing fibre reinforced concrete products and extensive Project Management. In the late 1980s he first encountered Value Engineering and quickly recognised its potential to address many of the key issues of successful project management and its close relationship with Managing Risk.

Through the UK's Institute of Value Management (IVM) he played a key role in developing the practice of Value Management in UK. He represented the UK in the development of the European Standard in Value Management (now the British Standard BS EN 12973:2000). On behalf of the IVM, he directed the development the European Training and Certification System. This is now the adopted system throughout Europe and South East Asia.

In 1997 he joined Davis Langdon as their Partner responsible for developing Value and Risk Management services. Davis Langdon (www.davislangdon.com) is a global project and cost consultancy, whose mission is to add value and reduce risk for its clients in the construction industry. It employs around 3000 professional staff in offices around the world.

This broad base provided the opportunity to build one of the largest Risk and Value Management consultancies in the world, as well as developing and embedding a Value and Risk culture in a large, international organisation.

Michael is a Member of the Institution of Civil Engineers and Fellow of the Institute of Value Management.

# Acknowledgements

I would like to thank the many people who have helped Michael Dallas and the Chartered Institute of Building (CIOB) with the production of this book, but in particular, the working group and their organisations for their valuable review of Michael's text; Saleem Akram, CIOB Professional and Technical Development Director for managing the review process; Rosemary Elder, CIOB R&D Manager and Sue Dennison, Executive Assistant, for assembling and coordinating the working group; Davis Langdon for their encouragement and generous support and the assistance of Beatrice Esprit, Jill Kirk and Emma Webb in editing the text and diagrams.

I would also like to thank the authors of the guest chapters for their contributions.

Finally, I would like to record my appreciation of Michael Dallas for writing this book, without whom there would be no publication.

**Chris Blythe**
Chief Executive
CIOB

## Working group members

**Gavin Maxwell-Hart** BSc, CEng, FICE, FIHT, MCIArb — Institution of Civil Engineers

**Roger Waterhouse** MSc, FCIOB, FRICS, MSIB, FAPM — Association for Project Management, Royal Institution of Chartered Surveyors

**John Campbell** BSc (Hons), ArchDip AA, RIBA — Royal Institute of British Architects

**Stephanie Clackworthy** FCMA, MBA, PVM, TVM — Institute of Value Management

**Emma Major** — Institute of Value Management

**Saleem Akram** BEng (Civil),
MSc (CM), FIE, MASCE, MAPM,
MACostE, FCIOB

**Professor John Bennett** DSc,
FRICS

**Richard Biggs** MSc, FCIOB,
MAPM, MCMI

**Alastair Blyth** BSc (Hons),
DipArch, MA, RIBA

**Ian Caldwell** BSc, BArch, RIBA,
ARIAS, MIMgt

**John Douglas** FIDM, FRSA       Englemere Limited

**Rosemary Elder** BSc, PGCE

**Barry Jones** FCIOB

**Arnab Mukherjee** BEng, MSc (CM)

**Kristina Smith** BSc, Eng

**David Trench** CBE, FAPM, FCMI

**Professor Graham Winch** PhD,
MCIOB, MAPM

**David Woolven** MSc, FCIOB

**Nausheen Shah** (Student) BSc
(Civil), MSc (CM)

# 1 Key features and benefits

## 1.1 Why successful projects need value and risk management

Value and risk management enables organisations to succeed in the delivery of ambitious projects by defining their desired outcomes and then exercising processes that maximise value and minimise uncertainty. This applies equally to strategy and business change projects as it does to those in the built environment. A successful outcome requires that the value to the business is maximised through the delivery of a facility that gives them the benefits they need at a price they can afford at the time when they need it and to a quality that fulfils their expectations. It requires that the outcome is clearly defined and communicated to those who deliver it (the project team). It also requires effective delivery processes that minimise the impact of the unexpected and uncertainties.

Value management provides an effective process to maximise value in line with the owners' and end users' requirements, and fulfils the first of these requirements. Risk management fulfils the second requirement as part of effective project management, by providing a process for managing risk. Both processes should be applied on every significant construction project. This does not always happen. This chapter explores why this is and provides arguments for their systematic use.

## 1.2 Delivering success

The effective, formalised processes of value and risk management enhance the chances of project success for minimal outlay.

The Eden Project (see Figure 1.1), in Cornwall, inspired by Tim Smit and designed by Sir Nicholas Grimshaw & Partners is a unique destination. It combines educational facilities, research and stunning tourist attractions within the overarching theme of sustainability. It is probably the most successful of all the millennium projects in the United Kingdom. It used value and risk management to great effect in overcoming seemingly impossible obstacles in fund raising, design and construction, to open ahead of schedule within the budget and exceeding expectations.

## *Value management*

At the outset of a project, value management provides an exceptionally powerful way of exploring the client's needs in-depth by addressing inconsistencies and expressing these in a language that all parties, whether technically informed or new to the construction industry, can understand. This results in the following benefits:

1.  It defines what the owners and end users mean by value, and provides the basis for making decisions, throughout the project, on the basis of value. It provides a means for optimising the balance between differing stakeholders' needs.

2.  It provides the basis for clear briefs that reflect the client's priorities and expectations, expressed in a language that all can understand. This improves communication between all stakeholders so that each of them can understand and respect other's constraints and requirements.

3.  It ensures that the project is the most cost-effective way of delivering the business benefits and provides a basis for refining the business case. It addresses both the monetary and non-monetary benefits.

4.  It supports good design through improved communications, mutual learning and enhanced team working, leading to better technical solutions with enhanced performance and quality, where it matters. The methods encourage challenging the status quo and developing innovative design solutions.

5.  It provides a way of measuring value, taking into account non-monetary benefits and demonstrating that value for money has been achieved.

(a)

(b)

**Figure 1.1** The Eden Project – the hugely popular humid tropics biome

**Figure 1.1** Continued

## *Risk management*

In his report – *Trusting the Team* – proposing improvements to the construction industry, Sir Michael Latham stated 'No construction project is risk free. Risk can be managed, minimised, shared, transferred or accepted. It cannot be ignored.'

It is necessary to take risk if one is to maximise the benefits (or value) of an organisation. The first and major benefit of risk management, therefore, is that it enables senior management to embark upon projects in the full knowledge that they will be able to control risk and thereby maximise their rewards.

Engaging in fire fighting, while it may be exciting is not efficient. It concentrates management's attention on day-to-day matters while diverting attention from the wider issues. Risk management as described in this book helps the team to concentrate on the big issues and manage these in an orderly way.

A formal risk management process delivers the following benefits for the project team:

1.   It requires that the management infrastructure is in place to deliver successful outcomes. This includes setting clear, realistic and achievable project objectives from the outset.

2.   It establishes the risk profile of the project, enabling appropriate allocation of risk, so that the party best placed to manage it has the responsibility for doing so. Risk allocation is a key component of contract documentation.

3.   It allows the team to manage risk effectively, concentrate resources on the things that really matter, resulting in risk reduction as the project proceeds. It also enables them to capitalise on opportunities revealed through the use of the process.

4.   It improves confidence that the project will be delivered to the owners' and end users' expectations, within the constraints of time and cost and to the required quality.

5.   Quantification of risk assists management in the task of raising the necessary funds and, later, controlling the project by judicious application and draw down of the risk allowances. Where the project forms part of a larger portfolio of projects, it enables the transfer of risk allowance from one project to another.

6.  It provides a mechanism for reporting risk on a regular basis to the appropriate levels of management, escalating severe risks in an orderly manner to obtain direction from the highest levels.

## Integrated value and risk management

Combining value and risk management studies into a single integrated process enables the project team to enjoy all the benefits outlined above. In addition the combination delivers other benefits.

1.  The processes are complementary enabling each to augment the other.

2.  Both disciplines require a deep understanding of the project. The discipline of acquiring this deep understanding helps the team to make better-informed decisions and improves communications between them. Improved communications lead to improved understanding. The process is cyclical.

3.  The framework provided by the value and risk management processes improves communication between the members of the team and external stakeholders, so that they can arrive at better solutions faster.

4.  The records of value and risk management studies provide an audit trail to demonstrate that activities that add value and reduce risk were actively managed. This enables third parties to understand the basis for decisions that have been made.

5.  The existence of good records provides the basis for learning from past experience and continuously improving performance.

## Alignment with construction industry improvement initiatives

In recent years there has been an increased requirement for good corporate governance. This has filtered down to corporate activities, including project management. Clients and customers are becoming more demanding. At the same time, construction projects are becoming ever more complex. In these circumstances, the formal processes of value and risk management when rigorously applied provide a structured route for the team to arrive at optimum solutions and demonstrate good project governance.

Recent initiatives to improve performance in the construction industry have focused mainly on the delivery processes. These are embodied in the key performance indicators (KPIs) that are supported by the Department of Trade and Industry (DTI) and Constructing Excellence. Important though these are in improving the project delivery process, less emphasis has been placed on ensuring that the outcome results in a product that meets, or exceeds, the expectations of the organisation that commissions the project. Value and risk management is focused firmly on delivering successful outcomes.

The contribution that they make in delivering successful outcomes is acknowledged and promoted in the Organisation for Government Commerce (OGC) Achieving Excellence Guidelines. These identify value and risk management as key contributors to successful project delivery.

The National Audit Office, responsible for checking that taxpayers get value for money, has, with the support of the Council for Architecture in the Built Environment (CABE), published guidance to its auditors under the title 'Getting Value for Money in Construction Projects through Design'. This publication is based on value management principles.

## 1.3 Summary

Formal value and risk management maximise value, reduce uncertainty and maximise return on investment.

Between them they enable the project team to put in place the conditions for successful projects:

*Clarity of purpose.* Value management provides the means to express the long-term project objectives in terms of the benefits expected by the owner and end users, unambiguously and in language that is clear to all. This includes a clear statement of business needs and a definition of value to provide the basis for the project and design briefs. Risk management enables the team to minimise uncertainty in delivering the benefits.

*People.* The processes used encourage clear leadership by providing a structure for making decisions to increase value and reduce risk. At the same time, people are encouraged to build a culture in which they work together towards a common purpose. Roles and responsibilities are clearly defined, assisting the selection of the right people for the job. External and internal stakeholders are consulted so that their views are taken into account and reconciled.

*Communication.* The processes involve extensive consultation and workshops, encouraging effective communications through agreed channels. Good communications improve decision-making and problem resolution.

*Realistic and affordable budgets.* Systematic application helps clients to set realistic, affordable and achievable budgets based upon the expected benefits flowing from the project. Demonstration of budget realism, together with an appreciation of the risks involved, contribute to the development of a sound business case. Active management of value and risk enable the supply chain to deliver the required quality within the budget.

*Appropriate procurement.* Appreciation of the project risk profile and allocation of risk to the appropriate parties best able to manage them provides the basis for selecting the most appropriate procurement strategy.

*Achievable programme.* Application of the processes from project inception encourages the adoption of realistic programmes with enough time allowed for design and adequate preparations before construction is committed, enhancing the ability to predict completion dates.

*Efficient delivery.* Optimisation of delivery processes leads to improved performance during design and construction and a minimisation of abortive time and waste. Coupled with the clear articulation of the required outcome, this maximises efficiency in terms of cost, time and quality.

# **2** Principles of value management

## Summary

- This chapter describes the essential attributes of value management and outlines the history of its evolution from value analysis.

- The concept of value (for money) and how value management focuses on defining the project outcome before exploring how to get there are discussed.

- It describes how the value management family comprises three distinct components, related to the project stages:

  - Inception to feasibility      Value articulation and project definition

  - Design and construction      Optimisation of benefits and costs

  - Commissioning and use      Learning lessons and performance optimisation

- Each of these components is outlined in this chapter and described in detail in later chapters.

- It shows how the focus of value management evolves through the life of a project and how the need for continuous improvement can lead to new projects.

- It outlines a generic process that ensures that the most appropriate methods are used for the task in hand.

- The common confusion between value and cost and its relevance to balancing investment with benefits is explored, emphasising the need to identify the main whole-life cost drivers and to take cashflow into account when assessing affordability.

■ The value of time is demonstrated by reference to the cost–time–quality triangle. Its relevance to projects, particularly those in which the outcome is a strong revenue stream, is explained.

■ A distinction is drawn between quality as conformity with specification and customer focused quality (giving the customer what he needs).

■ The end of the chapter describes the use of value profiling to assess and measure value, taking into account factors other than cost, which are crucial to project success.

## 2.1 Essential attributes

Value management concerns defining what those who commission projects mean by value, in terms of the expected benefits and then applying certain structured techniques, throughout the life of the project, to maximise the delivery of the benefits, while minimising the use of resources.

The British Standard describes it as a management approach to help maximise the overall performance of an organisation. The standard recognises that this is achieved through the application to projects at strategic, tactical or operational level.

'Definitions' abound; however, in the authors view it is not practical to capture all the nuances of value management in one or two succinct sentences, any more than it is possible to 'define' project management in a few words.

Value management is about more than simple cost reduction. While it is a very effective way to do this, it does much more. Essentially, value management:

■ Provides a means to define projects clearly and unambiguously in terms of the client's and end users' long-term business needs.

■ Supports crucial decision-making based upon maximising value.

■ Encourages innovative solutions.

■ Facilitates the optimisation of the balance among initial design costs and construction costs, running and maintenance costs, and the costs of operating the business conducted within the facility.

■ Provides a means of measuring value and demonstrating that optimum value for money (VfM) has been achieved.

The techniques used to deliver the above benefits are explained in detail in Chapter 10.

## 2.2 A brief history

The techniques that underpin value management were first formalised by an American, Larry Miles, while working with General Electric in 1947, shortly after World War II. Production of certain components was constrained by shortages of the materials that traditionally had been used in their manufacture. Miles was charged with finding ways to overcome the material shortages. The approach he adopted was, at the time novel. Instead of asking the question 'how can we find alternative sources of materials' he asked, 'what function does this component perform and how else can we perform that function'. This approach opened the way for innovative designs that resulted in superior products that were cheaper to make. Miles's focus on function (what things must do) was effectively a focus on the outcome of the manufacturing process rather than the process itself. Fulfilling the essential functions of the product under investigation, with the use of minimum resources, remains at the heart of value management today.

It was not long before people realised that not only did the technique provide a way to substitute alternative materials, it was also an excellent way to reduce costs while maintaining the all-important necessary functionality. In many cases, cost reduction became the main focus of value analysis, often at the expense of functionality. This abuse of the technique founded the myth that value management is primarily a method to reduce costs. One of the main reasons for writing this book is to dispel this myth.

Miles named his technique value analysis, a term that is still in use today. It was so successful that its use soon became widespread. In less than 10 years, the technique was adopted by the US Department of Defense (DoD) to improve VfM in the delivery of its very extensive construction programme. From the outset, the programme delivered cost savings of millions of dollars, averaging around 10% across all projects.

There is an apocryphal story that, when value analysis was adopted by the US DoD there were no vacancies for analysts but there were for engineers. Thus instead of recruiting value analysts the US DoD recruited value engineers. Whatever the truth of this tale, it was at about this time that the term value engineering (commonly shortened to VE) was born.

Although its use was widespread in the United States, it took another 30 years before the technique was used for the first time in the United Kingdom, by the American company Xerox on its UK headquarters building in 1983.

One of the reasons for the slow take-up in the United Kingdom was that the American approach to VE required a separate VE team to 'audit' the incumbent design team's proposals. The incumbent design team was then expected to implement the VE team's proposals. This opened up all sorts of problems relating to design responsibility and professional rivalry, leading, unsurprisingly, to strong resistance from the incumbent team. In Chapter 5 we explain how the modern approach is to work *with* the incumbent teams to ensure that all proposals for improvement have their support. While many designers are still suspicious of VE, this consensual approach has overcome much of the resistance described above.

Over the years the focus of value management (as it has come to be known) has matured. Initially, the emphasis shifted from material selection to cost reduction. The next evolution came about with the growth of the quality movement in the 1980s to enhancing quality as defined by conformity with specification. Gradually there was a growing realisation that this was not enough. Products had to respond to the needs of the customer. Emphasis therefore shifted to a more rounded appreciation of value, embracing performance, cost, time, knowledge and brand. At the same time the scope of studies expanded to embrace processes as well as products. In the context of construction, recent focus in best practice in value management has moved onto the theme of articulating and measuring value in a way that the project team can respond with the most effective solutions. At the same time the objects under study have evolved from products to include processes (see Figure 2.1). Each of these stages was built on previous experience resulting in the complex appreciation of value we see today.

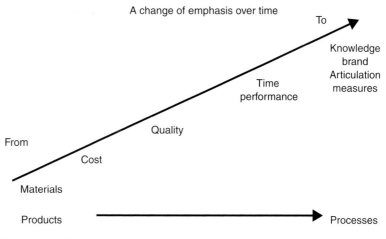

**Figure 2.1** The evolution of value management

## 2.3 Language

All management processes develop their own language. Value management is no exception. However, despite efforts to standardise the use of such language, there will always be those who use terms differently.

The British Standard defines certain terms for use in value management. Practitioners who have been brought up under the American system, as promoted by the Society of American Value Engineers (SAVE), use slightly different terms compared with practitioners in the United Kingdom. The terms used in this book are summarised in the Glossary in Appendix B.

Because some people may use different terms, study leaders should become acquainted themselves with the terminology used in the organisation within which they are working so that he uses the terms that fit with their culture.

## 2.4 Concepts

To understand value management completely the people involved should understand the concept of value. The Oxford English Dictionary definition of the word 'value' is 'the worth, desirability or utility of a thing, or the qualities on which these depend'. This is the meaning of the word as used in this book. It is not the same as 'values' which the dictionary defines as 'one's principles or standards; one's judgement of what is valuable in life'. In the context of

$$\text{Value} \quad \propto \quad \frac{\text{Benefits Delivered}}{\text{Resources Used}}$$

**Figure 2.2** The value ratio

construction, the term value is generally taken to mean the balance between how well the building satisfies the owner's expectations and the sacrifices, in terms of resources used, he must make in order to get it. The ratio between benefits delivered and resources used is referred to as the *Value Ratio* (Figure 2.2). Because the use of resources often comes down to money, the ratio is often referred to as VfM.

In construction projects, the resources used comprise land costs, materials, time and labour. All of these can be measured unambiguously. Measurement of the benefits delivered may be more complex. Financial benefits may be assessed unambiguously. Non-tangible benefits, such as aesthetic considerations, require a different approach. The techniques of function analysis (describing things by what they do, rather than what they are) provide a powerful way to do this. The methods are discussed in more detail in Chapters 6 and 10.

Value management is concerned with defining what is needed in a project to deliver the expected benefits and then ensuring that these things are implemented effectively, economically and efficiently, using the minimum resources. Value management, as the name implies, is an approach aimed at maximising value on a particular project. It is not simply about reducing costs.

Value is subjective, with different people applying different criteria to assess whether they are getting good value. For example, the developer of a block of flats may seek to spend as little as possible on the building while charging the highest rent that the market will bear. The tenant, on the other hand, seeks spacious (therefore expensive) accommodation at minimum rent. If he perceives that the rent is too high for the space offered he will not rent it. The developer thus loses value. Value management seeks to achieve the optimum balance between these two extremes. It provides the means to reconcile the different stakeholders' needs.

Maximising value is about delivering maximum benefits while consuming minimum resources. In short, value management is about delivering more of the *right* things for less resource.

## *Focus on outcomes*

When he was chairman of British Airways, Sir Colin Marshall remarked that, 'you read a book from beginning to end. You run a business the opposite way – you start at the end and then do everything you must do to reach it'. This echoes one of the core concepts of value management.

A key differentiator between value management and many other processes is that it focuses on the expected outcome from a project. Only once the outcome is clearly defined and understood does it address how it will be delivered. The desired outcome from a project is represented in a statement of the project objectives, expressed in terms of the expected benefits to the business. These are linked through the value drivers to the design intent. Later, as the project evolves, these relate directly to the design solutions and what is built (see the value cascade, Figure 6.3).

In any value management study it is important never to lose sight of the desired outcomes of the project. Frequently, simple cost cutting falls into this trap. It reduces the costs at the expense of benefits. Value will only increase if the value ratio increases. Value will be added if both the value ratio *and* the amount of benefits increase.

## 2.5 The family

In much the same way as project management encompasses many activities, value management is a generic term embracing a host of methods and techniques to assist in maximising value. Different methods and approaches are applied at different stages in project development. There are three distinct phases:

| Project stage | Focus of activity |
|---|---|
| Inception to feasibility | Value articulation and project definition |
| Design and construction | Optimisation of benefits and costs |
| Commissioning and use | Learning lessons and performance optimisation |

These phases lead to the following study types which are described in Chapter 8.2:

| No. | Type | Typical question answered |
|---|---|---|
| $V_M0$ | Need verification | Is this the right project |
| $V_M1$ | Project definition | What are the project objectives? |
| $V_M2$ | Brief development | What is the best option? |
| $V_M3$ | Value engineering | Is this the most cost-effective solution? |
| $V_M4$ | Handover review | Did we achieve our expectations? |
| $V_M5$ | Post-occupancy/ use review | Is the business sustainable? |

## *Value articulation and project definition*

The use of functions to describe the benefits expected of a project enable the team to build direct links between the project objectives and the design solutions that later get built. These functions are often referred to as value drivers, since they encapsulate what creates value for the client and end users. The techniques for doing this are described in Chapters 6 and 10. Essentially, by helping the project team to build a function model of the project helps them do three things:

First they can agree on a clear description of what the project must achieve in terms of the benefits expected by the client and the end users.

Second they break this down into simple functional statements that describe the levels and quality of benefits.

Third they break these down into clear statements that communicate to the design team those things that they must take into account when they develop their designs.

Taken together, these describe the value that is expected at the outcome of the project and inform the development of the brief. They also provide the criteria upon which decisions should be based.

## *Optimisation of benefits and costs*

Using the function model described above, the design team can develop designs that accurately reflect what the client and end users expect. However, we have seen that satisfying the benefits in full represents only part of the optimisation of value represented by the value ratio. To deliver good VfM the team must make best use of the resources that are available.

This is where techniques such as VE are useful. This technique builds upon the function model developed in the project definition stage to the point where it identifies the proposed building elements that are linked with each function. By adding the estimated costs of providing the elements that contribute to each function, the team builds up a picture of how much each function costs. By comparing this with the importance of each function they can assess where they are getting good VfM and where they are not.

The team generates ideas for performing the functions in different ways, starting with those that appear to offer lower VfM. It evaluates the relative merit of the ideas and develops those with most promise into detailed proposals for improvement. Finally, it submits its recommendations, based upon the proposals that it has developed, to the decision-makers who will decide which to include.

The technique is quick and very effective, provided the team applies it rigorously and is not tempted to take short cuts. It is described in more detail in Chapter 8.

## *Learning lessons and performance optimisation*

Once the construction stage of the project is completed the team should conduct a project review. This provides the opportunity to check with the team and the building's users whether the full benefits that were defined at the outset of the project and predicted in the VE proposals were realised. The review can explore what went well and what could have been improved and provide the opportunity particularly successful generic value management proposals to be 'banked' for use on future projects.

Effective ways of conducting project reviews and embedding the lessons learnt from them are discussed in Chapter 8.

## 2.6 Value management through the project cycle

The techniques employed will vary depending on the stage in the project. In the early stages, the first step is to verify that a construction project is the best way to go about delivering the benefits that are sought from it. The next stage is about defining a project that will deliver those benefits. It is only later in the project, when

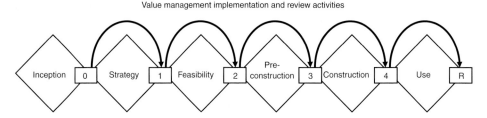

**Figure 2.3** Value management through the project cycle

a design has been developed, that one uses VE (described above). Chapter 8 contains a full description of the methods used at different stages of a project. The flowchart in Figure 2.3 summarises the process and its relationship with the risk management process (described in the next chapter).

The format for each study will usually follow the stages of a typical VE study, tailored to suit the project stage and the study objectives. These are summarised below and described in more detail in Chapter 8.

| | |
|---|---|
| Stage 1 | Preparation |
| Stage 2 | Workshop briefing, function analysis (and option selection, if appropriate) |
| Stage 3 | Creativity |
| Stage 4 | Evaluation |
| Stage 5 | Development |
| Stage 6 | Presentation and reporting |
| Stage 7 | Implementation |

These are described in more detail in the section titled, 'Value Engineering' in Chapter 8.

## *The value cycle*

In order to remain competitive all businesses must continuously improve their competitiveness. This includes measures to make better use of their assets. For owner-occupiers, one of their greatest assets is their building stock. Value management may be applied not just to new building projects. Once a building is in use, changes in the business environment, improving technology, changes in working practices, as well as the passage of time take their toll on the value of the asset to the business. Value management may be used to great effect to identify and then optimise delivery of asset improvement projects. These may range from building replacement to improving working processes. The diagram in Figure 2.4 sets out the essential steps in the value cycle.

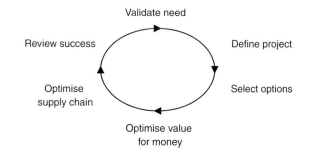

**Figure 2.4** The value cycle

## 2.7 The generic process

It is important to distinguish between the value management process, encompassing all the activities in a continuing programme of value-adding activities and the VE job plan described in Chapter 8. There is frequent confusion on this point. VE is a specific and very effective technique to optimise the cost-effectiveness of a design. At different stages in the project, other techniques (see Chapter 8) may be used to enhance value. The process shown in Figure 2.5, adapted from the British Standard BS EN 12973:2000, proposes a simple study plan to ensure that the most appropriate techniques are used and that the skills are in place to use them effectively.

## *Defining the objectives*

The first task is to define the objectives and targets that the value management study is to achieve. Study objectives differ from the project objectives since they describe the intended *study* outcome. They will vary as the project develops. In the earliest stage of a project, the inception, the study objectives are likely to focus on verifying that a construction project is needed and that there is not another, better, way to deliver the required benefits. At the concept stage of the project, the objectives will concentrate on evolving a clear definition of the project to inform the project brief. During the feasibility stage, the objective will be to inform the selection

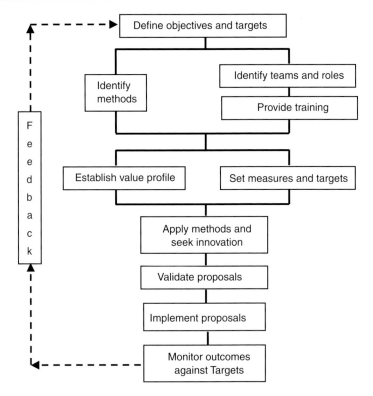

**Figure 2.5** The value management study plan

between the various options that may be put forward by the design team and informing the development of the design brief. Once a design solution has been decided and developed in sufficient detail (usually Stage C or D in the RIBA plan of work, or 30–40% design completion) the objectives will focus on improving the design solutions to deliver better quality and optimise cost and time. These study types, and others, are described in greater detail in Chapter 8.

## Selecting teams and roles

Only once the objectives of the study have been identified, can the study leader, with the assistance of the client and other project team members, identify who should be involved, what roles they will play and what are the most appropriate value management methods to apply in order to achieve the objectives. If it is found that members of the identified team lack some of the skills to apply the methods effectively, it may be appropriate to offer them training to improve their skill base to a satisfactory level.

## Identifying value drivers (functions)

The next stage is to identify the main functions that need to be fulfilled by the project in order that it may deliver the benefits expected of it (these may be described as value drivers for the project). The methods used for function analysis are described in detail in Chapter 8. It is good practice at this stage to establish the hierarchy of importance between the main value drivers, thus establishing the value profile for the project. This will be useful to assist decision-making and in selecting options.

## Target setting

Now that the study objectives, the principal functions, (or value drivers) and the value profile are understood, it is appropriate to quantify the objectives by setting targets expressed in practical metrics. Targets may be expressed in terms of the value drivers and thus provide a direct way to measure value improvements not constrained to time and cost alone.

## Apply the methods

The next stages in the study are to apply the selected methods to generate proposals to meet the study objectives. These proposals should be worked up in sufficient detail to inform management decisions (see Chapter 8).

## Validation and Implementation

Once a proposal has been accepted, in principle, by the management, the project team will need to validate that it is sound and that the assumptions upon which it is based are valid. Only then should it be included in the agreed action plan for implementing the value management proposals. It is vital at this stage to agree an implementation plan for tracking progress, monitoring outcomes and comparing these with the targets originally set. This is a frequent failure in many value management plans. If, having undertaken a study, all the good ideas and proposals that came out of it are not rigorously tracked to the end of the project, much benefit will be lost.

## Feedback

Lessons learned from one study should be captured (e.g. by way of a post-occupation project review) so that they may improve performance on subsequent studies.

## 2.8 Value or cost – balancing benefits and investment

A common misconception in the construction industry is that value is synonymous with cost. This may be brought about by sloppy use of language rather than a fundamental misunderstanding of the meaning of the terms. Frequently, the press will refer to the value of a project as being the capital cost. They may then go on to describe how the project team through skilful use of various management techniques (including VE) succeeded in making large reductions in value! Clearly, such references are to cost not to value. Few clients would be pleased if the value of their project had, in fact, reduced.

Ask many people in the construction industry the value of a brick, and they will respond that bricks cost, say, £250 per thousand. That is simply the cost of the brick. Its value lies in the fact that we can build a wall from it, support other parts of the structure on it and that, if well made, it can impart an aesthetic value to the project. In other words, the value of the brick lies not in what it is, but what it can do for the benefit of the building. For some this represents a paradigm shift in their thinking – from what things are to what things do. The principle is illustrated in Figure 2.6 and applies to all

The paradigm shift

**From:** **what things are** – a Brick

**To:** **what things do** –
supports building;
excludes weather;
enhances appearance;
retains heat;
protects against cold;
minimises maintenance

A brick has no value until it does something useful

**Figure 2.6** A Paradigm shift in thinking

elements used in a building. Their value lies in what they bring to the project, not what they are. Unless an element in the project can be linked through the value tree to one or more of the value drivers, it brings no value but simply adds to the costs. This is a strong argument for its elimination. Before eliminating it, however, the team should ensure that they have not failed to identify a necessary function that is performed by the orphan element.

In the value ratio, cost is but *one* of the resources used in delivering the benefits, albeit a very important one. The confusion between value and cost is one of the reasons why VE, or worse, value management, is viewed by many as simply a way to reduce costs. VE *is* a very effective way in which to reduce costs but *not* at the expense of benefits.

## Cost drivers

The principle cost drivers to creating a building are likely to be land acquisition and the main construction elements, such as substructures, structure, envelope, M&E services and fit out. Understanding these and how they contribute to the value drivers is critical to the use of VE to reduce costs.

Creating the building is, however, only a small part of the cost of owning and operating it. Our industry frequently fails to consider both initial and whole-life costs. In their paper to the Royal Society of Engineers, the authors, Evans Harriott Haste and Jones, suggested that a typical ratio between construction costs, building operating costs and business operating costs can be in the order of 1 to 5 to 200 (this ratio is heavily dependent upon the use of the building) (see Figure 2.7).

The figures demonstrate that the long-term costs of running a building and of conducting business within it far exceed the initial cost of constructing it. The long-term costs are therefore the greater cost driver. It is essential that the long-term or whole-life costs attaching to a building be fully understood. VE proposals should always be compared on a whole-life cost basis.

The case is illustrated by a bank that undertook a programme to modernise and re-design its branches. Post-occupancy reviews indicated that working conditions improved so much that staff productivity increased by 23%. The resulting reductions in staff costs effectively paid for the refurbished branches in less than a year.

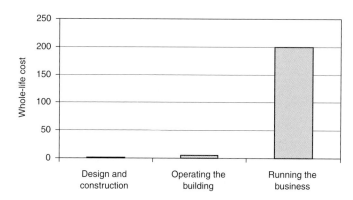

**Figure 2.7** The whole-life cost of a building

## Whole-life costs

In order to calculate whole-life costs for a building it is necessary to understand the main cost drivers throughout the life of the asset. Cost headings are generally gathered under four main categories:

1.  Initial costs – these are one-off costs incurred only when a building is constructed (it is usual also to include under this category disposal costs which are the one-off costs or income generated when a building is disposed).

2.  Periodic costs – these are costs which may be incurred from time to time to replace various components within the building or to refurbish it.

3.  Annual costs – these are the costs incurred on a day-to-day basis, the total of which can be assessed over a full operating year. They will include staff, rates, energy, cleaning and general maintenance costs. For an owner-occupier they will include the costs of operating the business.

4.  Income – this will include rental income and, for an owner-occupier, income derived from the activity undertaken within the building. Income may also be divided into initial, annual, periodic and disposal cashflows.

The technique of whole-life costing is discussed in Chapter 9.

## Cashflow and affordability

A further VfM consideration is a project's affordability. If the most economical solution involves spending more money than is available in the short term, it will be unaffordable.

In these circumstances it is essential that the optimum solutions be matched with the available cash.

> The consolidation of the Defence Research Agency (now DERA) from around 50 sites around Britain to just half a dozen involved a comprehensive value management programme to ensure that the proposals represented best VfM. One of the constraints within which the project teams had to operate was the strict cashflow profile laid down by HM Treasury at the outset of the programme. This meant that it was necessary to strike a balance between the optimum programme for construction efficiency and the availability of cash to pay for it.

## 2.9 The value of time

In many organisations the value of time is no less important than that given to cost. For example, in the retail trade, opening a new store in the run up to Christmas is of paramount importance since any overrun will put the profits to be made during the peak Christmas period in jeopardy.

It has long been recognised in project management that there is a strong relationship between time, cost and quality. This is normally represented by a triangle with time, cost and quality at its corners. On any project either time, cost or quality may be of supreme importance. Attempting to give all three equal importance can risk achieving none of the targets. If a time deadline must be met and if there is a finite budget for the work which cannot be breached, it is likely that quality will suffer (see Figure 2.8). Likewise, if quality is paramount and costs are fixed, it is likely that it will take longer than

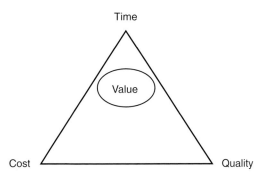

**Figure 2.8** Time, cost, quality triangle with an emphasis on time

expected to complete the project. One of the aims in value management is to achieve the optimum balance between time, cost and quality, recognising that all three are not necessarily obtainable in full.

Being first in the market place with a new product can be worth a huge premium. A food company had developed a new additive, which simplified the making of cakes as well as enhancing their health characteristics. They wished to construct a factory to supply the product in commercial quantities in time for an international food fair, which was scheduled to take place within 6 months.

The construction cost estimate was about 20% higher than planned. A VE study not only reduced the construction cost to within acceptable limits but also identified ways to accelerate the construction programme to enable completion and commissioning within the very tight timeline.

There is generally an optimum timeline for constructing a building. Reductions in construction time come at a premium in cost and, sometimes, a reduction in quality. Many value management studies overlook the value of time. When establishing the study objectives it is worth considering whether the root cause of the problem is time rather than cost.

A good example of the value of time occurred during a study into loss of market share of a compressor manufacturer. At the outset of the study, management assumed that they were being undercut on cost grounds by their competitors or that their compressors were less efficient. Detailed analysis showed that the root cause of their problems lay in the fact that, by the time they had compiled all the data needed to respond to enquiries, several weeks had elapsed. The competitors were much more responsive and were gaining market share. A detailed value management study focusing on their enquiry response processes succeeded in reducing their response time from several weeks to less than 1 day! Here time was the problem, not cost or machine efficiency.

## 2.10 Quality

The Gucci family motto is said to be that 'quality is remembered long after the price has been forgotten'. Iconic buildings, such as the Sydney Opera House, are seldom built down to a price but the value of their outstanding qualities is often underestimated during their construction.

Peter Drucker has been quoted as saying: 'There is nothing so useless as doing efficiently that which should not be done at all.' Building efficiently is no guarantee that the end product is suitable for its intended purpose.

Quality in buildings is often expressed as conformity to specification. The specification writer will carefully describe the attributes of an element in the construction and the contractor will endeavour to match this specification. Provided the specification is matched or exceeded, the contractor will have been deemed to have delivered quality. This is fine so far as it goes. However, using this definition it is possible to achieve excellent quality but fail to deliver something that works for the building owner.

A more customer-focused definition of quality is provided by the value driver model referred to earlier. The value drivers describe the required quality of the project in terms of benefits to the owner and/or users. Choosing the right language for the terms in the project objectives and the description of the value drivers provides a very clear way to describe the level of quality expected by the client.

For example, use of the term 'world class' in describing a feature means that you will not find anything better anywhere in the world. This is a tall order and can be very expensive. The term may be used rather glibly when, in fact, the client is seeking something which is very good and better than comparable facilities in that country or region.

Best in class or best in Birmingham is a much more realistic expectation and may be perfectly adequate to deliver the competitive advantage that is needed. World class implies an emphasis right at the extreme of the quality corner of the time, cost and quality triangle (Figure 2.9); it is, therefore, likely to cost a lot and takes a long time to deliver.

There are, of course, plenty of instances where world class is precisely what the client means. In these cases the client must understand that it may be necessary to pay a premium to achieve it. The development of a value profile provides an excellent way to describe the level of quality required across the project as well as the relative importance of the components within it. It is therefore an excellent tool to inform the design brief represented.

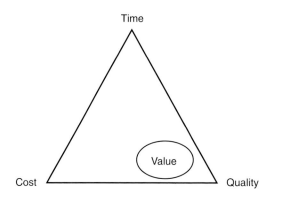

**Figure 2.9** Time, cost, quality triangle with an emphasis on quality

## *Use of value propositions to define specification*

In Chapter 6, we introduce the concepts of value propositions and the balanced score card. The former concept helps organisations select and communicate their preferred market orientation in terms of value as perceived by their customers. The latter provides a model upon which to measure their past and predicted performance in terms of the financial parameters, their customers and their business processes.

It is a relatively simple matter to group the value drivers in the same categories as used in an organisations' value proposition and the balanced score card, thus clearly relating project value drivers to business objectives. This principle is shown in Table 2.1.

**Table 2.1** Value profile aligned with balanced score card measures

| Business measure | Value driver | Importance weight % | Attainment (1–10) | Value score |
|---|---|---|---|---|
| Financial | Maximise sale value | 45 | 6 | 270 |
| Customers | Attract tenants | 10 | 3 | 30 |
| Enabling | Obtain approvals | 35 | 8 | 280 |
| Processes | Manage effectively | 10 | 7 | 70 |
| Total value index (out of 1000) | | | | 650 |

## 2.11 Measuring value

No management process can be said to be truly effective unless it is possible to measure its effectiveness. Wherever possible such measures should be objective and unambiguous. In some cases there may be an unavoidable element of subjectivity. In such cases, it is necessary to build a consensus, for example, by taking a series of observations from a number of stakeholders or by undertaking a survey. As a direct result of the recommendations of Re-thinking Construction, Constructing Excellence (formerly the Construction Best Practice Movement) has developed a series of key performance indicators (KPIs) for use in the construction industry. The main KPIs for all types of construction are:

- Client satisfaction – product (what is built)
- Client satisfaction – service (how well the team performed)
- Defects
- Predictability – cost
- Construction time
- Predictability – time
- Construction cost
- Safety
- Productivity
- Profitability

Most of the emphasis of these KPIs are on the *process* of delivering construction. By contrast, value and risk management focuses on project *outcomes* rather than the process for delivering project success. While the above KPIs are useful to improve construction processes, it is necessary to develop a method by which we can measure the value added to a project based on outcomes rather than process. In Section 2.7 we saw how the expected outcome of a project may be described through functions or value drivers. This concept can be utilised to measure the value delivered by a project.

In Section 6.4 we describe a table of nine value drivers which are generic to most building types. When drafting 'Getting value for money from construction projects through design', sponsored by the Commission for Architecture in the Built Environment (CABE) and the National Audit Office (NAO) distilled the number down to six key value drivers.

The six key value drivers adopted by CABE/NAO are:

1.  Achieve required financial performance (ensuring the building is affordable in whole-life terms).

2.  Manage the project effectively (minimising waste and maximising efficiency during construction).

3.  Maximise business efficiency (of activities conducted within the building when it is complete).

4.  Project the appropriate image (to the outside world as well as those who use or visit the building).

5.  Minimise building occupancy and maintenance costs (once the building is in use).

6.  Comply with third-party requirements (including legislation and health and safety).

Table 2.2 demonstrates how these may be used to assess the value for, say, a school. The method used is to develop what is known as a value profile for the project.

**Table 2.2** School value profile

| Value driver | Importance weight, % | Metric | Performance (on scale of 1–10) | Weighted value score |
|---|---|---|---|---|
| Financial | 20 | Cost | 3 | 60 |
| Manage project | 15 | KPI | 4 | 60 |
| Business efficiency | 30 | Output | 3 | 90 |
| Image | 10 | Survey | 6 | 60 |
| Occupation cost | 15 | Cost | 8 | 120 |
| Third-party requirements | 10 | Audit | 4 | 40 |
| Total value index | | | | 430 |

*Note:* Guide to total value index: 850 excellent; 750 good; 500 room for improvement; 350 requires improvement.

## Value profiles to measure value

The team first identifies the value drivers. Next it weights them to establish their relative importance. This creates the value profile. The project team agrees appropriate metrics for each of the value drivers. For each value driver the team establishes performance ranges from unacceptable to delight. The team then assesses current performance against each of the value drivers. Multiplying the performance (on a scale of 1–10) by the importance weighting (expressed as percentage) gives a value score for each value driver. Summing all the value scores yields an overall value index for the project. This process may be referred to by the acronym VAMP and is illustrated in Figure 2.10.

The higher the value index, the greater is the value, 1000 being the theoretical maximum. In practice, an index of 850 is excellent, while an index of 350 or less requires improvement.

We illustrate the application of the above process to a school building in Table 2.2, indicating how an organisation might apply the above principles to assess its value. The above method provides an effective way to measure value in an easy to understand quantified outcome, taking into account both monetary and non-monetary considerations.

In this example, a value index of only 430 indicates a clear room for improvement and signals the need for a value study.

Successive reassessments of the value index after each value study can give the project team a clear indication of how effective their

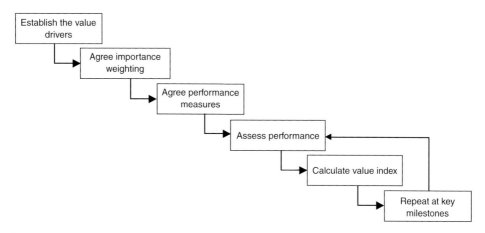

**Figure 2.10** Value measuring process (VAMP) summary

efforts have been and where additional effort is needed to improve value still further.

## *Setting metrics*

One of the more difficult tasks in using the methods described above is agreeing objective metrics for attributes that may be somewhat subjective. Ideally, a metric should be identified that cannot be influenced by the observer. For example, costs and quantities of materials can be estimated from the market and are therefore truly objective. Even relatively soft values can be the subject of objective metrics. For example, levels of satisfaction in a hospital outpatients department may be measured by the number of letters of complaint (or praise) received.

Where such objective metrics are not possible, it may be necessary to resort to surveys, ideally undertaken by independent observers.

# 3 Principles of risk management

## Summary

- This chapter first outlines the essential attributes of risk management and why it is needed in construction projects.

- It goes on to describe the evolution of risk analysis from gambling to present day requirements of corporate governance.

- In describing the underlying concepts of risk, this chapter outlines why it is necessary to gain a full understanding of the project before gathering information to assess risk and takes the reader through the key steps of the process.

- It explains that, in the context of construction projects, the risk family comprises three areas of risk: business, project and operational, and how these may arise at any time during the project.

- The chapter goes on to describe how the process is cyclical throughout the project and outlines the generic risk management process.

- It goes on to describe its use in construction projects, how risk may be measured both qualitatively and quantitatively, and introduces the reader to the estimation of cost-related risks and uncertainty. It outlines computer-based risk analysis.

- Next it introduces the concepts of contingency management and its application to programmes of projects.

- Finally, it explains why it is not possible to add time risks in the same way as cost risks and how this difficulty may be resolved by means of computer analysis.

## 3.1 Essential attributes

Construction risk management is concerned with improving confidence that the outcome of the project will deliver the business benefit expectations. It provides a structured way to capture the experience of the project delivery team to reduce the chances of things going wrong, take advantage of opportunities and increase their ability to make decisions in the best interests of the project.

Statistics published by Re-Thinking Construction in 2002 show that about 40% of projects are delivered late, 50% of projects overspend their budget and over 30% of projects fail to fulfil the expectations of the users. With statistics like these one must ask the question: 'Can I afford *not* to undertake formal risk management?'

The discipline of a formal approach is made all the more necessary in the absence of reliable historical data on the uncertainties faced in construction projects. This contrasts with the application of risk management in other sectors, for example, insurance or financial, where risk management is used to predict outcomes using actuarial databases build up over many years.

The Association for Project Management (APM) define risk as 'an uncertain event or set of circumstances that, should it occur, will have an effect on the achievement of one or more on the project's objectives'. The management of risk involves minimising these uncertainties. If the impact of the consequence of risk occurring has been overestimated, then reduction of uncertainty can present opportunities to improve the outcome.

The process of risk management can be quite straightforward but requires leadership, buy-in from the project team and rigour in its application. Like value management, it should be applied throughout the project cycle (see Figure 3.1).

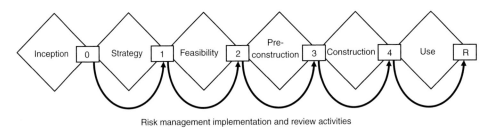

Risk management implementation and review activities

**Figure 3.1** Risk management through the project cycle

For the management of risk to be effective, it is important to keep it as straightforward as possible so that it remains manageable. Too much detail and complexity can make the task of active management of risk unwieldy with the result that the team loses interest in the process.

## 3.2 Evolution

The concepts of risk management have been around for hundreds, if not thousands, of years. It all started with gambling. Gambling is based on games of chance, where skill makes little or no difference. Gaming houses are able to make a profit because they can predict the chance of the gambler winning and tilt the balance in their favour. To make such predictions relies on a numbering system that facilitates calculation. Thus quantitative risk analysis had to wait until the evolution of the Arabic numbering system. In the seventeenth century a French nobleman, Chevalier de Mere, who was fond of gambling and mathematics, challenged the great mathematician Blaise Pascal to solve a two centuries old puzzle of how to divide the stakes between two players in an unfinished game of chance. His solution laid the basis for statistical calculations that underpin all quantitative risk assessment. Use of statistical formulae enabled astute businessmen to pitch the odds in their favour in various business transactions (whether these arose through gambling or through other business transactions such as insurance). Businesses such as insurance began to accumulate actuarial data to enable them to predict the likelihood of certain events and thus set premiums at appropriate levels to be profitable.

This has over the years led to the development of sophisticated risk management processes applied to business issues. It has been extensively used in organisations whose core business is dependent upon an accurate assessment of risks. Most prominent among these businesses are the financial sector and insurance companies. Actuarial data of this type is not, in the main, available in the construction industry, making accurate prediction of risk a particularly subjective and risky business. As a consequence, those organisations whose core business is not directly dependent upon risk tend to adopt a much more *ad hoc* approach.

Every business takes risks. If these are not managed effectively, whether through formal processes or not, they can find that trading conditions become so bad that it must cease business. Shareholders in businesses that have no formal risk management in place feel particularly aggrieved if the senior management and directors, in whom they had placed their trust, appear not to have managed their businesses professionally.

## *The Turnbull Report*

In 1999, the government commissioned Professor Turnbull to address this situation. His report recommended, amongst other things, that all businesses should have in place a robust process for managing risk. The principles set out in the Turnbull Report now set the standard for good corporate governance across all sectors, both public and private.

The report made the following recommendation:

> A company's objectives, its internal organisation and the environment which it operates in are continually evolving and as a result the risks it faces are constantly changing. A sound system of control therefore depends on a thorough and regular evaluation of the nature and extent of the risks to which the company is exposed. Since profits (or business results) are in part the reward for successful risk taking in business, the purpose of internal control is to control risk rather than eliminate it.

Raising the profile of risk management across business generally, combined with the high incidence of construction projects failing to deliver to expectation, has led to a rapid development of formal risk management processes in the construction industry during the second half of the twentieth century. One of the big influences was the development of the oil fields in the North Sea during the 1970s. Oil platforms were constructed on land and then towed out to be placed in several hundred feet of hostile North Sea during very short weather windows. The need to predict accurately the completion dates of a complex oil platform against these tight timelines acted as a catalyst for the development of formal risk management in construction. At the end of the twentieth century the British Standards Institution (BSI) published the Project Management Standard BS 6079 of which the third part relates to the management of risks. This document is not confined to construction projects and includes the management of risk in business-related projects. While an understanding of business risk is useful in successful project delivery it is not the subject of this book. The project management standard does, however, provide some very useful guidelines of best practice in the management of risk on construction projects. There is no British Standard relating to risk management in construction projects.

## 3.3 Language

The language used to describe risk management can vary from organisation to organisation. Consistent terminology is useful, but slavish adherence to terminology that may be recommended within any one methodology should not dictate the terms used in all circumstances. The study leader should adopt terms that are consistent with the language used in the organisation within which he is working, provided that these are, in turn, consistent with effective risk management. The terms outlined in the Glossary in Appendix B attempt to steer a middle course through the language used in differing guides and uses terms that are in common use in the construction industry. What is important is that those involved in the risk management process understand the terminology that is being used on their particular project.

## 3.4 Concepts

The saying 'nothing ventured, nothing gained' neatly captures the principle that taking risks is necessary in order to gain rewards. Thus project risk management is not about eliminating risk altogether but controlling the risks to which the organisation is exposed when undertaking a project from which defined benefits are expected.

In order to control anything we need to understand it. The Sage of Omaha, Warren Buffet, is probably the world's most successful investor. He is quoted as saying:

> Risk is not knowing what you are doing.

Mr Buffet has become extremely successful by making sure that he understands every aspect of what he is doing before committing on an investment.

The first job in risk management, therefore, is to understand the project in depth. The following paragraphs explain the underlying concepts of risk management by taking the reader through the essential steps of the process.

### *Understanding the project*

We list below some of the key questions that the risk management study leader should ask before embarking on any study.

Readers will note that these questions are the same and should be asked when embarking on a value management study. This is unsurprising since both must begin with a thorough understanding of the project.

- Why is the project being undertaken?

- What are its objectives?

- What are the benefits that are expected to flow from it?

- Who is involved?

- What are their interests?

- Within what business environment is it being undertaken?

- What are the financial parameters relating to this project?

- What is the timescale?

- What are the constraints?

- What are the factors that are critical to its success?

- What design solutions are being proposed?

- How much is it expected to cost?

- Is there enough money available to pay for it?

## Risks

Having gained answers to these questions the study leader is well on his way towards gaining a comprehensive understanding of the project. He can then begin to identify what could prevent the full realisation of benefits. These are the risks to the project.

## Opportunities

At the same time as identifying what creates uncertainty in the delivery of the benefits, one needs to understand the assumptions that have been made. These will normally have been embedded in the project business plan. Some of these could be overly cautious and, therefore, might actually turn out better than expected. These represent opportunities that can be exposed through risk management. Having identified any opportunities, it is wise to communicate them to the value management activities that are undertaken in parallel to risk management so that they can be maximised.

## Consequences

Having identified the risks, we need to understand what would happen should those risks occur. These are the consequences of a risk occurring and are the things that could cause damage. The risk itself can remain a threat throughout the project but never cause any damage (apart from sleepless nights).

## Impact

It is only when the risk occurs that it will have an impact on the project. The impact may be financial, it may be related to the duration of an event, or it may affect the quality of the product. It may, of course, affect all three. For each risk it is necessary to assess, qualitatively or quantitatively, the impact of its consequence on each of these dimensions.

## Likelihood

The final essential piece of information we need about a risk, in order to consider how we might manage it, is to understand how likely it is that it will occur. Again, this can be expressed qualitatively or quantitatively, qualitatively as 'likelihood' and quantitatively as a 'probability'. Where something that may affect the project outcome is certain to happen, it is referred to as an issue. Issues need to be addressed by the project team in a similar way to risks.

## Interrelationship of risks

We now have a much better understanding of risk to a project than before we conducted the above analysis. There is, however, one other important factor which is necessary to understand to enable effective management. That is interrelationship between certain risks. It is not uncommon for one risk to trigger another. For example, in a project where it is necessary to move occupants from one part of the building to another, to enable phased refurbishment works, a delay in one phase of the refurbishment work can disrupt a carefully structured decanting plan in another area of the building.

Such knock-on effects can transform a relatively minor event (e.g. delay in the redecoration of a single room) to a much more major event, for example, inability to open a hospital ward, because opening the hospital ward required moving some thing or some people into the room which was late in decoration. It is said that, in chaos theory, the flapping of a butterfly's wing in Brazil can trigger a hurricane in the Caribbean; there are similar considerations in risk management. While such risks cannot in themselves be managed, it is possible to put in place contingency plans to minimise the impact, should they occur.

> The redevelopment of a major hospital in southern England required the refurbishment of a six-storey block that had been built in the 1960s. To maintain the operation of the hospital it was not possible to vacate the block. It was therefore necessary to vacate parts of the building, refurbish the vacated parts and then move staff and/or patients into the refurbished parts so as to permit refurbishment of the next part. This required a detailed decanting plan. Any delay on one section under refurbishment would have a knock-on effect on the next part. Because timelines were tight, this represented a major risk to the redevelopment programme. Unfortunately, the service lines within the block had evolved in a random manner and no accurate plans existed of how to isolate the vacated areas without disrupting services to the areas that remained in use. The situation was compounded by the presence of uncharted asbestos in the ceilings. Detailed advance surveys were not possible since the areas to be surveyed were in constant use. Recognising that the existing decant plan represented too great a risk to the project overall, the team rescheduled it to include generous float and contingency plans.

## Escalation

In circumstances like these it is essential that the project team has in place a mechanism for reporting tasks to senior management. This is referred to as escalating a risk.

## 3.5 The family

Risks to construction projects may be grouped into three headings:

- Those that could affect the business of the organisation commissioning the project – these are termed business risks.

- Those that could affect the successful delivery of the project – generally termed project risks.

- Those that could affect the use of the built facility – generally known as operational risks.

Unlike the three components of value management, which are related to the stage in the project when a study takes place, the categories of risk are not time related. Operational risks may arise from decisions made during project inception, just as much as from decisions that are made later in the programme. This is one of the reasons for involving the planning supervisor at an early stage in the project.

Likewise, business risks can arise from events that occur during the project design and construction stages. This is why it is essential to have a risk reporting escalation procedure in place.

The fact that a facility is in use does not preclude circumstances giving rise to business risks – for example, if the facility does not fulfil the purpose for which it was intended.

## 3.6 Risk management cycle

Risk management should be viewed as a continuous activity throughout the development of a project, resulting in a steady reduction of the level of risk across the project as it proceeds. As a project evolves, some risks will be allocated to different parties in the project team, some will pass and no longer represent a threat and new risks will emerge. Some management actions will prove effective, others less so. Figure 3.2 outlines the key activities in this iterative process.

**Figure 3.2** The risk management cycle

## 3.7 A generic risk management process

The process for managing risk outlined above is summarised in the Project Management Standard, BS 6079-3 : 2000 and illustrated in Figure 6.11.

The first thing to do is to establish the context of the study. Understand the objectives of the project, how it fits with the business overall, its scope and other characteristics.

The process of identifying and analysing risks, outlined under Section 3.4 and the next section, allows the team to evaluate whether it is necessary to actively manage the risk and what management strategy should be adopted, this leads to identifying action owners and the risk management plan.

The next stage is to undertake those actions to treat the risk in the manner that has been agreed. At regular intervals the effectiveness of this treatment should be reviewed and if necessary adjusted. As the project proceeds new risks will require identification just as past risks can be deleted, all risks (both new and old) then require further analysis to check if their severity has changed – re-evaluation to amend the risk management plan if previous treatment was not effective, continuing treatment and further review. Mainly, risk management should be considered as a reiterative management cycle, repeated many times throughout the life of a project. There are, however, occasions such as project management gateways, when a more formal risk review may be required. The details of risk management studies consistent with the above cycle are discussed in Chapter 8.

## *Application to construction*

In a similar way that value management activities varies depending on the project stage at which they are applied, so will risk management (see Figure 3.2).

At inception, the focus of the study will be on what risks the project may pose to the business and strategic risks relating to the ability to conduct a successful project.

At strategy and feasibility stages, studies will focus on how risks are to be allocated between the parties, thus informing the procurement strategy, and the relative risks presented by different options.

As the project develops, in the pre-construction and construction stages, the focus will shift towards more technical matters, initially relating to construction and then risks to operating the completed facility. Throughout, the need to deliver a building that is safe and healthy both to construct and to operate, when completed, will be reviewed by the planning supervisor as required by law. The details of this process are not covered in this book since they are usually conducted as a separate activity stream. The planning supervisor should, however, be included in along with those who are consulted throughout the project risk management process.

Finally, in the handover and use stage, studies will look at operational issues and maintaining business continuity. This is shown diagrammatically in Figure 3.1.

The generic process comprises seven stages:

| | |
|---|---|
| Stage 1 | Preparation |
| Stage 2 | Identification |
| Stage 3 | Analysis |
| Stage 4 | Evaluation |
| Stage 5 | Treatment (or management) |
| Stage 6 | Presentation and reporting |
| Stage 7 | Implementation and review |

These stages are explained in more details in Chapter 8.

## 3.8 Risk to quality

There are two measures of quality:

1.    Conformity to specification

2.    Delivery of benefits.

There may be risks associated with both of these.

There is risk, for example, that the contractor may not have sufficient skills to conform precisely to the required specification required by the contract documents. This could be the consequence of a shortage of skilled labour or equipment, the need to hurry some activity, or as a result of an unrealistically tight specification laid down by the design team. All of the above consequences are relatively easy to predict and to remedy. They represent deviations from specification.

What tends to be less easy to predict or remedy are those risks which affect the delivery of benefits. These can arise due to shortcomings in the management infrastructure or the environment within which the project is undertaken. They may result from shortcomings in design. Risks arising from management shortcomings or the project environment are, by definition, likely to be less easily controllable by the project team. An effective approach for dealing with such risks is central in creating the conditions for a successful project and is discussed in Chapter 11.

## 3.9 Measuring risk

The experienced manager will draw on this experience to gauge the risk factor of something in order to determine the 'best' course of action. Intuition as a management tool is unscientific and subjective. It depends upon the individual's appetite for risk as well as his judgement. For repeatability, a more scientific method is needed.

The method most commonly used for measuring risk in the construction industry is based upon estimating the likelihood that a risk will occur and the impact of its consequence(s), should it occur. Risks may be assessed qualitatively or quantitatively depending upon the ultimate use of the results. If the data is to be used only to set up a system for managing risk, qualitative analysis is usually sufficient. If, however, the results are to inform a risk allowance, the measurement *must* be quantitative.

### *Qualitative assessment*

In the absence of actuarial data, the most common method used to assess the severity of risks is to use simple three- or five-point scales for impact and likelihood. Multiplying the likelihood by the impact gives the risk a rating (of severity).

Thus on the three-point scale, the most severe risks, high (likelihood)-high (impact), will rate 9 (3 × 3) and on the five-point scale the most severe risks will rate 25 (5 × 5) (see Figure 3.3).

One drawback with this system is that it does not differentiate between those risks that are highly likely to occur but will have low impact (it might rain) and those risks that are very unlikely to occur but would have disastrous impact if they did (the bridge could collapse). Both would score 3 (3 × 1) on the three-point scale or 5 (5 × 1) on the five-point scale.

Another method, advocated by the Institution of Civil Engineers in its publication 'Risk Analysis and Management for Projects' (RAMP). Here the risk that it might rain (highly likely but of negligible impact) would rate 16. The risk of a bridge collapsing (extremely unlikely but of disastrous impact) would rate 1000. The skewed

|   | Very low | Low | Medium | High | Very high |
|---|---|---|---|---|---|
| **Very high** | 5 | 10 | 15 | 20 | 25 |
| **High** | 4 | 8 | 12 | 16 | 20 |
| **Medium** | 3 | 6 | 9 | 12 | 15 |
| **Low** | 2 | 4 | 6 | 8 | 10 |
| **Very low** | 1 | 2 | 3 | 4 | 5 |

Impact (vertical axis) / Likelihood (horizontal axis)

**Figure 3.3** A 5 × 5 point risk analysis matrix

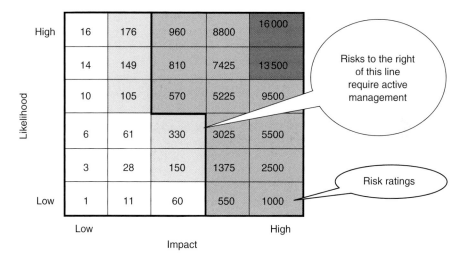

**Figure 3.4** Qualitative Assessment Matrix (based on RAMP)

matrix has a profound influence on the effort that will be put into managing a risk.

The risk matrices shown in Figures 3.3 and 3.4 are often referred to as 'heat maps' because the squares indicated the greatest risks (those that are highly likely and would have a severe impact) are coloured red, while lesser risks are coloured amber, yellow and green.

## Issues

When a perceived risk is certain to occur, it is called an issue. In management terms it should be treated exactly like any other risk but with a probability of 100%. If its rating is severe (i.e. it has a high impact) actions should be put in place to reduce its impact to below the acceptable threshold.

## Quantifying risks

Quantification of risks can lead to enhanced benefits to the project team. This may be done in terms of cost and/or time. Reasons for quantification could include:

■ When there is a need to report upwards or to third parties.

■ Where the project forms part of a larger programme of projects, enabling the transfer of risk allowance between projects.

■ To encourage people to follow through management actions.

■ Where clients require it as part of their standard procedures or have capped funds.

■ Where it is desirable to link contingency to risk, thus removing the expectation that it will be spent.

■ Where it is required or provides comfort to fund-raisers or other third parties.

Where quantification is required, a statistically sound method should be used. Suitable techniques are covered in Chapter 11.

## Estimation of cost of risks

Most risks are likely to have a direct or indirect impact on cost. Even if the direct consequence of the risk leads to a delay, its effects can often be translated into costs.

## Direct estimation

It is sometimes possible to estimate the cost impact of a risk occurring either as a single number or as a spread of numbers. Delay may have a direct cost impact if, for example, it delays or interrupts an income stream or increases overhead costs or preliminaries.

It may have no direct cost impact if it comprises a non-time-critical activity.

## Top-down risk estimation

The cost impact of more complex risks may not be so easy to calculate because of the difficulty in predicting the outcome of its consequences. For example, the potential delay caused in the decant programme described in Section 3.4 can be very difficult to predict. It will depend upon the circumstances and the consequences of related risks. All we know about this particular risk is that its impact is likely to be severe. In circumstances like this, it is often better to ascribe a range of costs to the qualitative assessment of the impact. This is generally expressed as a percentage sum of the total project cost (or, better, its net present value – NPV). This is explained in more detail in Chapter 11.

The 'top-down' method of estimating the cost impact, while it is quick, provides only a guideline to the likely costs incurred. It does, however, provide a way of quantifying the impacts of risk quickly and consistently across the whole project.

A central government department needed to undertake a comprehensive refurbishment of its central London headquarters. The project was complex, not just in construction terms, but also with respect to the logistics to ensure uninterrupted service. The risk management team identified a large number of risks and assessed them qualitatively, using the RAMP guidelines. They then agreed cost and probability ranges against each of the qualitative assessments and, using the simple spreadsheet analysis of the Central Limit Theorem, described in Chapter 11, calculated an appropriate risk allowance. The project team used the quantified risk register as a central tool in managing the project to a successful conclusion.

To calculate a more accurate estimate for such a risk, the team could work out all the possible permutations and combinations of the impacts of all the related risks using one of the techniques described in Chapter 11. This would be very time consuming and might not be any more accurate than the top-down method described above. It is necessary to apply judgement as to whether it is worth the extra effort of gaining a potentially more accurate estimate of the impact of risks.

Even where costs are relatively simple to estimate, they remain just that – estimates. It is important that the team remembers this when calculating risk allowances.

The uncertainties of cost estimation are a particular danger when applying computer analysis. Here a series of estimates may be converted in the computer to what appears to be a very accurate number. This principle is commonly known as the garbage in, gospel out (GIGO) principle and can confer an unrealistic sense of accuracy.

## Cost uncertainty

Most cost estimates are built up using estimated quantities and rates, based upon experience, to arrive at an estimated cost for the item in question, or element. There may be uncertainties in both the quantities and the rates used in compiling the estimate. It is one of the skills of the quantity surveyor to predict the most likely estimate for the element. Having built up an estimate of the construction costs of the building in this way, the level of uncertainty is usually expressed as a contingency sum.

A more scientific way to express the uncertainties of estimating is to ascribe a range of numbers to each estimate and analyse the likely outcomes using a computer (see below).

## Quantification of risk by computer analysis

There are many computer programmes available in the current market for predicting the outcome of risks occurring. Each has their strengths and weaknesses. All base their synthesis of the outcome of a project on the principle of running the project many times (e.g. 1000 iterations) while injecting each and every risk according to the probability and impact ranges established during the risk analysis. In all cases, the accuracy of the outputs (see Figure 3.5) are only as good as the information on which the analysis is based.

The total risk allowance generated by the methods outlined above is often significantly more than the level of contingency applied in traditional cost estimation. This is because the calculations are based on the *unmanaged* severity assessments, whereas the contingency assumes that risks *will be* managed effectively. In addition, a comprehensive register may include some risks, which are not usually articulated.

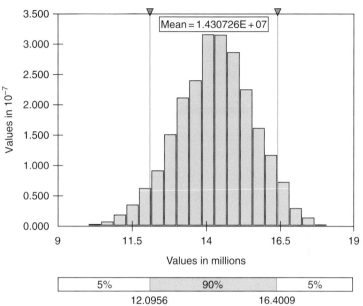

Distribution for MC/AA146

**Figure 3.5** A typical computer generated cost risk histogram

## 3.10 Contingency management

Risk allowances that are based upon a risk register contain direct links to actual events. This means that, if an event that contributes to the risk allowance passes without incurring cost or incurs a lower cost than has been estimated, the project manager can reallocate the amount saved to other projects or other cost centres within the same project. This process is known as contingency management. If the project forms part of a large programme of other projects, this provides a significant source of additional funding. For example, if risk allowances averaging 5% are released as soon as their outcome is known, an additional 1 project in 20 in the rest of the programme can proceed.

## 3.11 Time risk

Although many risks that relate to an activity's duration can result in a direct or indirect cost impact, they may also have an impact on the overall project duration.

Whereas cost risks may be summed across the whole risk register to inform the calculation of the risk allowance, it is not possible to

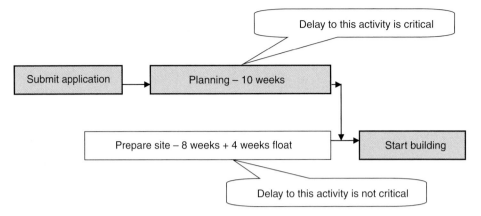

**Figure 3.6** Illustration of critical path analysis

simply add up the estimated time impacts (expressed as delays or accelerations) to arrive at the overall impact on the duration of the project.

To understand the reason for this, it is necessary to give an outline description of the critical path analysis method of estimating programme durations. Very simply, each activity throughout a project has a likely duration and is linked to other activities. The sum of all activities and their relationships to one another will determine the overall project duration. The process for estimating the overall project duration is known as critical path analysis as shown in Figure 3.6. There are many commercially available programmes offered to undertake this task.

The following narrative explains how this works. Let us assume that we estimate that it will take 10 weeks to obtain planning permission. This period cannot commence until we have submitted our application (this represents a logic link to a proceeding activity). We cannot start to build until we have received the planning consent (a logic link to the succeeding activity).

Critical path analysis as shown in Figure 3.6 allows us to build up a complete picture of the whole project using such linked activities and then calculate the shortest likely duration. Many activities are not critical to the overall project duration. If they take a little bit longer or a little bit less they will not affect it. These activities contain 'float', meaning there is time to spare and they are not critical. Other activities, however, are critical to the overall project duration and any delay in these will increase it. These activities have no float and are on the 'critical path'.

Thus the delay in one activity may or may not delay the project overall. Only if it lies on the critical path will a delay in an activity extend the overall duration. If a non-critical activity is delayed sufficiently, the delay may transform it into a critical activity and delay the project overall.

In order to estimate the overall time impact of risks occurring, it is necessary to identify which activities will be affected, estimate the impact on the duration for each of those activities and then analyse their effect on the overall project duration. The only practical way to do this is to run a computer analysis based on Monte Carlo simulation, linked to the critical path analysis. This technique is explained in more detail in Chapter 11 and should only be undertaken by someone who is skilled and experienced in its use.

# 4 An integrated approach to value and risk management

## Summary

- This chapter describes why value and risk management should not be viewed as separate activities but incorporated as an integrated part of project management. Each is complementary to the other in contributing to project success. Value management maximises value while risk management prevents value from being eroded or destroyed.

- It describes how the processes align well with each other and how they may be combined into a single integrated process.

- It goes on to outline the key stages in any project and when value and risk management can make an effective contribution to decision-making. It aligns the different study types with public sector gateway reviews and different procurement routes.

- The chapter underlines the importance of understanding the client's business and describes the roles of value and risk management in PFI projects, where a good understanding of the client's business is essential.

- It describes how studies should be planned at the outset of a project and integrated into a programme of activities to add value and control risk.

- It explains how studies at the beginning of any project provide an excellent way in which to launch the project and convey a sound understanding of the project imperatives to all project team members.

- It includes a table that outlines the focus of different study types, aligning value and risk management activities and summarising the intended outcomes at each of the project stages.

- The chapter goes on to list the critical success factors for a value and risk management programme and outlines a common framework for developing their practice within an organisation, in a way that develops the appropriate culture and delivers results.

- It gives guidance on the level of activity warranted by different sizes of project (Figure 4.9).

## 4.1 Why integrate?

To optimise value on a project the author believes that it is essential that the team actively manages both value and risk. There is little point in going to great lengths to maximise the value of a project if significant risks materialise which impair its delivery, thereby destroying value. A project in which all risk is avoided is unlikely to maximise value.

It is necessary to take risks to maximise value. Formal processes, rigorously applied, provide a structured route for the team to control risk effectively.

It is common practice to treat value and risk management as separate processes and many publications make this distinction. While the processes (as outlined in Chapters 2 and 3) may differ in detail, the recurring theme in this book is to encourage the consideration of both risk and value management simultaneously throughout the project (For an integrated approach to value and risk management see Figure 4.1).

Figure 4.2 illustrates how the integrated process comprises a series of joined up studies interspersed by separate streams of value and risk activities.

### *Value and risk are complementary*

Both risk and value management are needed to maximise the chances of a project's success. The reason for this lies in the different but complementary objectives of each discipline.

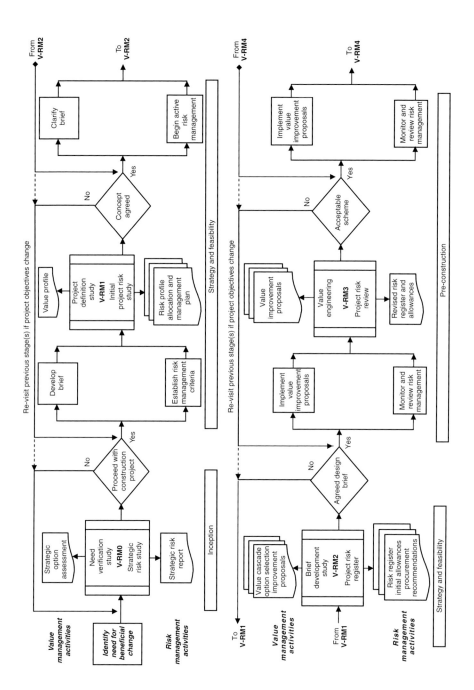

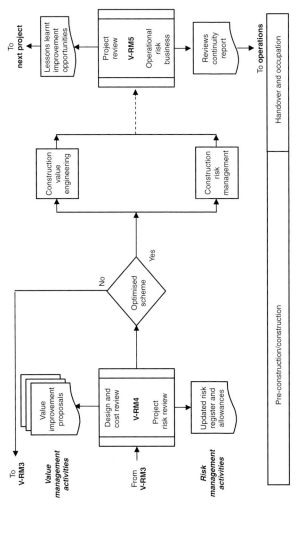

**Figure 4.1** Integrated value and risk management flowchart

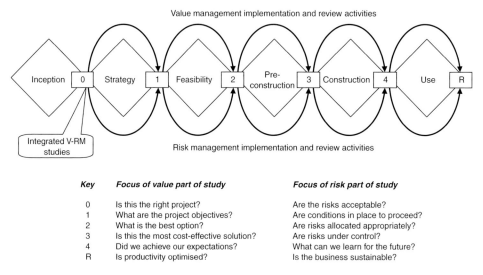

Value management implementation and review activities

| Inception | 0 | Strategy | 1 | Feasibility | 2 | Pre-construction | 3 | Construction | 4 | Use | R |

Integrated V-RM studies

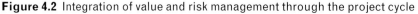

Risk management implementation and review activities

| Key | Focus of value part of study | Focus of risk part of study |
|-----|------------------------------|------------------------------|
| 0 | Is this the right project? | Are the risks acceptable? |
| 1 | What are the project objectives? | Are conditions in place to proceed? |
| 2 | What is the best option? | Are risks allocated appropriately? |
| 3 | Is this the most cost-effective solution? | Are risks under control? |
| 4 | Did we achieve our expectations? | What can we learn for the future? |
| R | Is productivity optimised? | Is the business sustainable? |

**Figure 4.2** Integration of value and risk management through the project cycle

Value management is concerned with clearly articulating what value means for the client and end users and then maximising its delivery by creating a clear link from the project objectives, via the design solutions, to the end product.

Risk management is concerned with identifying uncertainties in achieving the desired objectives (thus eroding value) and putting in place management action to avoid or minimise them.

Value is maximised using value management. Uncertainty and consequent value erosion is minimised using risk management.

## Relationship of risk with value drivers

In Chapter 3, we referred to the need to gain a full understanding of a project before the outset of any risk management. In Chapter 2, we described how building up a value driver model provided an excellent way to build up the team's understanding of the project by linking the expected benefits, expressed through value drivers to its design. This approach is just as effective to inform the team about managing risk. Since risk management is about reducing the uncertainty of delivering the expected benefits, a thorough understanding of the project value drivers is essential. The greatest risk on any project is that the expected benefits (as expressed through the value drivers) are not delivered in full.

## *Similarities in the processes*

While the processes may differ in detail, they have the following similarities:

■ The preparation stage, to understand the project and the issues relating to it.

■ The requirement for consultation with, and involvement of, the main stakeholders.

■ The use of facilitated workshops involving a balance of stake-holders, disciplines and characters.

■ The development of proposals to improve the project and management actions to implement them.

■ The need for an explicit implementation plan.

■ The written record, or report on the outcome, providing a clear audit trail.

■ The need for regular reviews to monitor implementation and report progress.

## 4.2 The integrated process

Combining the two processes within a single study is therefore logical and practical. The format for a typical combined study is shown in Table 4.1.

(Note that the Stages A–H in Table 4.1 refer to the stages in the integrated value and risk study, not the RIBA plan of work. The Stages 1–7 for risk and value refer to the individual study stages introduced in Section 2.6 for value and 3.7 for risk. These are described in more detail in Chapter 8.)

The progress of implementation and management actions should be reviewed by a responsible person on a regular basis and reported in the regular project reports. At this stage, between formal studies, the activities resulting from the combined value and risk study are likely to be conducted and reported separately. This is because

**Table 4.1** Integrated value and risk management

| Value (see Chapter 2) | Value and risk | Risk (see Chapter 3) |
|---|---|---|
| Stage 1 – preparation | Stage A – preparation and briefing | Stage 1 – preparation and Stage 2 – identification |
| Stage 2 – workshop briefing, function analysis | Stage B – workshop briefing, function analysis | |
| | Stage C – review risk register | Stage 3 – analysis and Stage 4 – evaluation (qualitative) |
| Stage 3 – creativity | Stage D – generate improvement ideas and management actions | Stage 5 – treatment (or management) planning |
| Stage 4 – evaluation | Stage E – evaluate ideas and select for development | |
| Stage 5 – development | Stage F – develop proposals and quantify risk allowance | Stage 4 – evaluation (quantification) |
| Stage 6 – presentation and reporting | Stage G – present recommendations and compile report | Stage 6 – presentation and reporting |
| Stage 7 – implementation and reviews | Stage H – implement proposals and actions, conduct regular reviews | Stage 7 – implementation and reviews |

different people within the project team may be responsible for conducting them.

## 4.3 Timing

### *Risk and value management plans*

If the project team is to maximise value and control risk it is necessary to have effective value and risk management implementation plans in place. These are discussed in more detail in Chapter 9. Essentially, they should show individuals' responsibilities, how they communicate with one another, how proposals to improve value or reduce risk will be implemented, by whom and within what timescales. They will also set out the timetable for regular reviews and formal studies.

Value and risk management studies should be planned from the outset of any project. They should not be regarded as sticking plaster remedies to dig the team out of a hole when things have

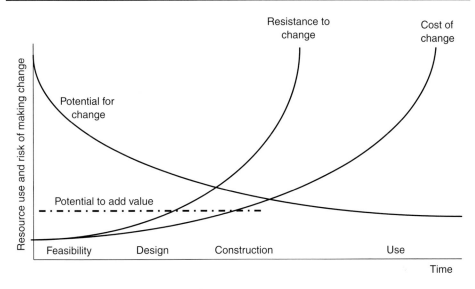

**Figure 4.3** Opportunities reduce with time

gone wrong. Sadly, this is often the way they are regarded, with the result that so-called value management exercises become little more than emergency cost cutting and risk studies are reduced to fire-fighting plans. The earlier that risk and value management activities are planned in the project lifecycle, the better will be the outcome, see Figure 4.3.

## Opportunities for change

It is well known that opportunities for making project changes reduce as the project develops. This is because the more decisions that are made and signed off, the more constrained the project is against change. For example, once detailed planning permission has been granted there is little opportunity for making fundamental changes to a proposed building.

## Cost of change

Likewise, the costs of making change will increase as the project progresses. The more advanced the design, the more complex it is to incorporate changes; this is due to the sunk-costs of abortive work as well as potential loss and expense claims by members of the project delivery team, if changes are made to signed off or contracted work.

It is commonly stated that 80% of the whole-life cost of a project is committed once the concept for the project has been agreed; a principle was confirmed based on surveys undertaken by the Construction Industry Institute in America in the mid-1980s.

## Resistance to change

One must not forget the human factor. The longer people have been working on a particular project, the more wedded they become to the plans that they have developed, and the less they will want to make changes. The project team therefore will often resist and oppose changes. Another reason relates to the way in which the services of the project team are procured. Nowadays, most designers and contractors are engaged on a fixed price to undertake a defined amount of work. Often they have won the commission in the face of competition with competitive bids. Any changes, therefore, will erode their profit margins, leading to claims or additional money and/or time. Where fees are related to the outturn cost of the project, changes that result in increases in outturn cost are unlikely to be resisted but there is a significant risk that the project may not deliver value for money.

## The Project duration, from inception to delivery of benefits

Traditionally, in the construction industry, a project is seen as the construction activity alone, beginning with an instruction from an owner and ending when the completed building is handed over for use. From the owner's perspective, constructing the building is a small component of a much broader project to bring about a benefit to a business. Until the construction is complete, occupied and working, it brings the owner no benefit whatsoever. It is simply an expense. A construction project should include the entire cycle, from realisation that something is needed in order to bring about beneficial change, to using the construction for the purpose for which it was commissioned. Indeed, on a whole-life basis, the project is not complete until disposal of the asset after use, a view that should be reflected in the business case. To the owner, the cashflow through the building project is represented by Figure 4.4.

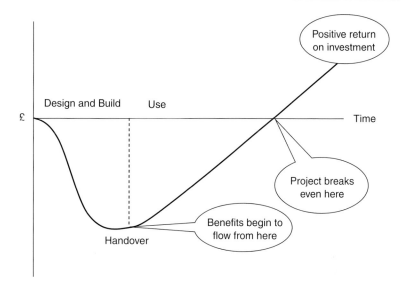

**Figure 4.4** Cashflow through the building project stages

Realisation that the client gains no benefit from a building until after it is in use is an important step in putting oneself in the client's shoes and delivering best value with minimum uncertainty (risk).

## 4.4 Project stages and study types

At key milestones it is advisable to conduct formal combined value and risk studies, typically comprising Stages A–G, as described in Section 4.2.

Although they may have a similar structure, each of the studies will have a different objective, linked to the stage in the project. The integrated value and risk management programme for a major project can be expressed by Figure 4.5.

### *The generic value and risk model*

This represents a generic model, aligning each study with a defined milestone or project stage. The end of each stage represents a gateway or decision point at which the management team will consider whether conditions are right to proceed to the next stage. Value and risk management studies are an excellent way of informing these decisions. Figure 4.5 illustrates different types of value management and risk management studies which might be undertaken at different milestones within a project.

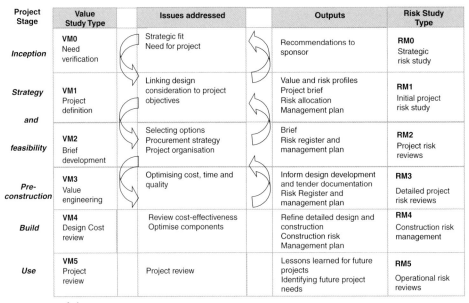

| Project Stage | Value Study Type | | Issues addressed | | Outputs | Risk Study Type |
|---|---|---|---|---|---|---|
| Inception | VM0 Need verification | | Strategic fit Need for project | | Recommendations to sponsor | RM0 Strategic risk study |
| Strategy and | VM1 Project definition | | Linking design consideration to project objectives | | Value and risk profiles Project brief Risk allocation Management plan | RM1 Initial project risk study |
| feasibility | VM2 Brief development | | Selecting options Procurement strategy Project organisation | | Brief Risk register and management plan | RM2 Project risk reviews |
| Pre-construction | VM3 Value engineering | | Optimising cost, time and quality | | Inform design development and tender documentation Risk Register and management plan | RM3 Detailed project risk reviews |
| Build | VM4 Design Cost review | | Review cost-effectiveness Optimise components | | Refine detailed design and construction Construction risk Management plan | RM4 Construction risk management |
| Use | VM5 Project review | | Project review | | Lessons learned for future projects Identifying future project needs | RM5 Operational risk reviews |

⟨ ⟩ *Arrows thus indicate potential reiterations which may be necessary if circumstances require strategic changes to the project*

**Figure 4.5** Milestones for Integrated value and risk management reviews

## Migration of study objectives

To reflect project progress, the objectives of value and risk studies change as the project moves from stage to stage. In the early stages, the emphasis is on ensuring that the project fits with the strategic needs of the business. Later studies are designed to inform the briefs to ensure that project teams develop solutions that deliver client and user expectations. When the design is being developed the team concentrates on ensuring effective project delivery by maximising value for money and controlling project risk. When a building is handed over for use, it is good practice to undertake a review to understand the effectiveness of the value and risk processes during delivery and learn lessons from them for future projects. Once the facility is in use, the users may wish to review the productivity of the operations from time to time. Such reviews can provide the catalysts for future projects.

## Iteration

The evolution of most development and construction projects is, by their nature, an iterative process. It requires constant juggling between the demands of the business, all the influential stake-holders, including the end user, the owner, the delivery team, the authorities and external pressure groups.

Until now, we have described the project stages as though they are a linear progression from inception to use. In reality, this is seldom the case. Circumstances can change. Changes can arise due to influences from the environment within which the project is taking place. Business strategy may change, enforcing changes on the project. Sometimes, change is needed because earlier project stages have not been rigorous. It may be necessary to go back a stage and review decisions that have been made earlier. Such itera-tion is indicated in Figure 4.5 by the curved arrows. The study leader should establish, during his preparations, whether decisions that have been made are still valid, consistent with the project objectives and based on sound foundations.

In the late 1990s the London International Financial Futures and Options Exchange (LIFFE) planned to build a new, modern head-quarters in the city of London in order to consolidate its position as world leader in its field.

The new centre was planned around the traditional method of trading, involving face to face bidding on an open trading hall.

At about the same time electronic trading was coming of age as an effective way of doing business. As the new technology developed so the directors of LIFFE realised that the future lay, not in grander trading halls, but in electronic trading. The way of doing business, open outcry, had changed. The plans for the new facility were abandoned.

## Project stages

Different sectors and disciplines tend to use different terms to describe the stages of a construction project. Much will depend upon the procurement route adopted, the lead discipline and the way in which the project is managed. What is important, however, is that each stage covers a definable set of activities towards the progress of a project. Each stage may include several milestones against which progress may be measured. Table 4.2 indicates the

**Table 4.2** Project stages in relation to common procurement stages

| Stage | VM study type | Study name: possible outputs | RM study type | Study name: possible outputs | Accelerating change stage | OGC gateway stage | RIBA stage | PFI stage |
|---|---|---|---|---|---|---|---|---|
| Inception | VM0 | **Need verification:** recommendations to management board | RM0 | **Strategic risk study:** recommendations to sponsor | 1 Verification of need | 0 Strategic assessment | A | Strategic outline case |
| Strategy and feasibility | VM1 | **Project definition:** information on which to develop project brief | RM1 | **Initial project risk study:** high level risk allocation and management plan | 2 Assessment of option | 1 Business justification | B | ISOP |
| | VM2 | **Brief development:** information on which to develop design brief | RM2 | **Project risk reviews:** updated risk register | 3 Develop procurement strategy | 2 Procurement method and sources of supply | C | ITN |
| Pre-construction | VM3 | **Value Engineering:** information to develop detailed design and tender documentation | RM3 | **Detailed project risk reviews:** updated risk register | 4 Implement procurement strategy | 3 Confirm investment decision; 3A design complete | D | ITN-PB; BAFO |
| Construction | VM4 | **Design and cost review:** refine detailed design and construction; improve supply chain efficiency | RM4 | **Construction risk management:** construction risk; management plan; project recovery plans | 5 Project delivery | 4 Readiness for service | E–H | Post-contract capex |
| Use | VM5 | **Project review:** lessons learned for improving management of future projects; informing activities to improve operational performance; identifying future project needs | RM5 | **Operational risk reviews:** business continuity plan | 6 post project review | 5 In-service benefits | L | Post-contract opex |

different activities that may be undertaken by value and risk management programmes at various stages throughout a project, related to different stage description.

The correlation in table 4.2 below is not exact. The mismatches in timing are particularly apparent in the alignment with gateway reviews (see later in this section). Gateway reviews were originally introduced for non-construction projects (such as ICT and defence projects), undertaken by turnkey contractors who are charged with developing, designing and implementing the project.

Gateway 3, investment decision, marks the beginning of the design development stage for most construction projects. There are no gateways that coincide with the key stages of design development although gateway review 0 may be used to check alignment with objectives at the key milestones. Development and construction projects should include additional reviews after gateway 3 and before gateway 4, before design is complete. The studies at this stage are likely to be aimed at optimising value for money in the design solutions and minimising uncertainty and risk in their execution.

## Client decisions

We have seen that value and risk management studies provide an extremely useful tool to inform a client decision as well as provide a very effective way of briefing the project delivery team and informing the brief.

If the purpose of a study is to inform a client decision, then it should be held shortly before a client approval milestone. If the main purpose is to brief the team on the next stage of work, then it should be held shortly after the decision to proceed has been made. Sometimes, it may be helpful to hold two events, one just before a milestone, to inform the client decision and a second, shortly after the milestone in order to convey the implications of the decisions to the project team.

If the study is to inform a client decision, it must be completed in sufficient time before the gateway so that the team's recommendations can be validated, evaluated and incorporated into the report to the management.

## *Public sector procurement best practice*

The above guidelines on timing are entirely consistent with the best practice recommended by government as described in the OGC's (Office for Government Commerce) Achieving Excellence Briefing Papers No. 04 – Risk and Value Management. This paper recognises that 'risk and value management are interrelated tasks that should be carried out in parallel'. The briefing paper also gives detailed guidance on the alignment of value and risk management studies with the gateway reviews. We reproduce this in Table 4.3.

## 4.5 Understanding the client's business

Buildings are commissioned so that the owner derives a benefit to his business, not to provide those employed in the construction industry with a living! Frequently a client's core business has nothing to do with building. To the client, commissioning a building is a distraction. Moreover, it may be but one of several ways in which to gain the benefits. If we are to deliver a project that meets their expectations, it is essential that we understand something about his business and the benefits that he is seeking. We need to change our mindsets from thinking 'what kind of building does the client want' to 'what benefits does he need?' Assuming that the project inception process recommends that a building is the best way to provide that need our next thoughts should be 'in order to deliver the benefits, what kind of building does he need?'

## *Private finance initiative*

Understanding the client's business is one of the key requirements of the private finance initiative (PFI) and outsourcing. Another is to allocate risk to the parties that are best placed to manage it. A government department's core business is to deliver quality services to the public. It is *not* to build, maintain or operate buildings.

For this reason, a core principle of PFI is to outsource the provision of buildings to specialists in the private sector, whose business it is to build, maintain and operate buildings. They provide a facility that enables the client to conduct his core business with maximum efficiency. In order to do this, they must have a sound understanding of their client's business. Under PFI the owner provides a detailed

**Table 4.3** Value and risk management stages from OGC's achieving excellence guidelines

| | |
|---|---|
| Before Gate 0 | Value management to identify stakeholder needs, objectives and priorities Risk analysis (high level) of potential project options |
| Gate 0 | Strategic assessment |
| Before Gate 1 | High level risk assessment. Value management study to evaluate options that could meet user needs |
| Gate 1 | Business justification |
| Before Gate 2 | Value management to develop output-based specification, to refine and evaluate options that satisfy project brief and objectives Risk management to identify risks for each procurement option, cost of managing them (through avoidance, design/reduction, acceptance, share or transfer). Revise risk allowance |
| Gate 2 | Procurement strategy |
| Before Gate 3 | Value management to apply selection and award criteria Risk management – update risk register and revise base estimate and risk allowance |
| Gate 3 | Investment decision |
| Before decision point 1: outline design | Value engineering study to optimise while-life design quality and cost. Integrated project team to assess buildability of options Risk management – identify residual risks and continue to manage risks and risk allowance. Agree and implement collective risk management approach |
| Decision point 1 | Outline design |
| Before decision point 2: detailed design | Value engineering study to optimise whole-life design quality and cost. Integrated project team to assess buildability of design Risk management – identify residual risks and continue to manage risks and risk allowance. Continue to implement joint risk management approach |
| Decision point 2 | Detailed design |
| Before Gate 4 | Finalise design and start construction Risk management ongoing during construction Value engineering for detail of finishes, etc. |
| Gate 4 | Readiness for service (construction complete) |
| Before Gate 5 | Risk management ongoing (service contract management phase for PFI, prime contract or design and build/manage and/or operate) Value management review and feedback of lessons learned |
| Gate 5 | Benefits evaluation |

output specification describing what the facility must *do* to enable him to provide his service effectively. This is also a core principle of value management – to identify what things must do – and thus it provides a good way of developing the output specification.

Thus PFI procurement is significantly different from conventional construction procurement, in that the client limits his risk to preparing a detailed output specification; the successful contractor must then build and operate the facility (at his risk) for a defined period, usually about 30 years.

Design, build, finance and operate contracts (DBFO), are similar, except that, whereas in a PFI-output specification the contractor must deliver specified availability of the facility, in DBFO the contractor guarantees a specified level of service delivery. This difference may seem subtle but makes a significant difference to the performance requirements of the contractor. For example, in PFI, the contractor may undertake to keep rooms clean and tidy at all times. This means that if someone makes the room untidy at any time the contractor must immediately make it tidy again. By contrast, the DBFO contractor undertakes to clean a room once a week. If someone immediately dirties it the day after it is clean, they are under no obligation to clean it until the next scheduled cleaning date.

Under the PFI procurement route there are effectively three risk and value management processes. The first is undertaken by the client to maximise value and minimise risk pre-contract. This will include identifying which risks will be allocated to the private sector partner on signing the contract and those that will be retained by the client after the contract is let. The second is undertaken by the private sector partner (or contractor) to optimise their response to the output specification and minimise the risk of default. The third is undertaken by the client to control his residual risk exposure post-contract.

## *Function performance specification*

The output specification referred to above is, in essence, a form of function performance specification. The specification does not describe the elements or components of the building. It should simply describe the functionality required to enable the client to deliver the service without defining the solution. This is described in more detail in Chapter 9. The method enables a much more innovative approach by the contractor and his supply chain.

During the strategy and concept stages of a major sporting venue project, the client held an integrated value and risk study to help build an appropriate output specification. The team developed a detailed value profile for the project, comprising an agreed statement of the project objectives, the main value drivers, reflecting the requirements of the client and the end users and linked each of these to the considerations that the project delivery team should take into account. While developing the model, the team chose the adjectives used to describe the functionality required very carefully. The result was a clear statement reflecting not simply the functionality that was required, but also the levels of service and quality that were expected.

The capacity for this approach to invite innovation was nowhere better illustrated than in an invitation to tender for the exploitation of an oil and gas field under the North Sea. This comprised a plan of the area showing the location and output data for the trial wells together with the statement 'Maximise the Value of the Asset'. How this would be achieved was entirely up to the bidders.

## 4.6 Programmes of projects

A major construction programme may comprise a number of interrelated projects. The overall programme objective must be reflected in a consistency of the objectives for each of the projects within that programme. A value and risk study held at the outset of the project will identify the core value drivers that apply to all projects within the programme. These can then be used to provide the basis for the individual project value profiles. A study of this type is often referred to as an anchor study because it provides a firm point of reference to which all subsequent projects may be anchored (see Figure 4.6).

The case study below describes how value management helped to provide consistency across the entire programme.

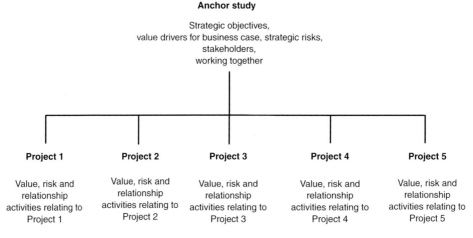

**Figure 4.6** Related projects within a programme

The Defence Research Agency (now DERA) used to occupy over 50 sites distributed around the United Kingdom undertaking all manner of research and prototype work on behalf of the Ministry of Defence. In order to reduce long-term costs, the ministry decided to consolidate operations into five centres. One of the challenges facing the project team was to achieve a consistency of accommodation across the whole estate, since the existing standards varied enormously. At the same time there was a paramount need to achieve value for money. To solve these problems, the project managers implemented an extensive value management programme. Consistency was achieved by convening an anchor workshop with senior management at which the team agreed with the high level value drivers for the whole project. These value drivers then formed the basis for building function cost models for each of the many business units to be accommodated in the five new centres. Value for money was assured by using these function cost models as the basis for value engineering studies on each of the sites.

## 4.7 Project launch studies

Another term that is used for a study at the outset of a project is a launch study. A combined value and risk study, together with activities that address how people are going to work well together, provides an excellent vehicle for undertaking. A successful launch, clearly defining the project value profile, identifying the key risks to the project success, establishing how people will work together and setting up active programmes for managing all of these things is crucial to the ongoing health of the project.

A progressive UK hospital planned a joint venture with an adjacent university to provide an integrated education centre in order to inspire learning. Hospitals have different priorities to universities and these were reflected in tensions in agreeing on the brief for the centre. Which views should prevail? Who should run it? How to reconcile the different (and equally important) positions? A project launch workshop involved all the key stakeholders and explored areas where there was full agreement, identified areas where there were differences and encouraged building a consensus of agreement. The outcome was an agreed and affordable brief for a ground breaking facility which is now built and successful in use.

## Stakeholder conferencing

Large programmes or projects involve many people, particularly in their early stages. These people may or may not be active contributors to the project. They can, however, be profoundly influenced by it and/or have the ability to impose their influence on it. These people are commonly referred to as stakeholders. Those who actively contribute to the project include the client body, advisors, the project delivery team, their supply chains and the users. We call these internal stakeholders. External stakeholders may comprise members of the public, the authorities, pressure groups and others. For success all these stakeholders should be consulted. The value and risk management methods described above provide a very effective way of conducting a stakeholder conference with large groups of stakeholders. The objective is to arrive at a consensus for the most advantageous project to meet the required objectives. Details of the method are given in Chapter 7.

Over many years various plans have been put forward to improve the setting of the unique monument of Stonehenge, culminating in proposals to bury the A303 from view and restore the immediate surroundings of the stones to something close to how they were when they were conceived. The project involved many stakeholders, each with their individual agendas. These included the Highways Agency (congestion free, safe roads), English Heritage (preserving ancient monuments and remains), National Trust (landowners and giving public accessibility), English Nature (preserving flora and fauna), Environment Agency (clean rivers and aquifers), Local Authorities (politics, planning and residents' interests), Farmers (making a living off the land) and others. To be a success the project must reconcile all these diverse interests and more.

A stakeholder conference involving about 30 stakeholders addressed four topics:

1. Identifying objectives that were common to all stakeholders.

2. Identifying the factors that were critical to the success of the project (the value drivers).

3. Exploring structures for roles, responsibilities and communications.

4. Gaining a common understanding of value, quality and risk.

The outcome was a broad degree of consensus on all topics, and resulted in an agreed basis for developing the project brief and a clear plan for defining roles, responsibilities and communications strategy.

## 4.8 Summary of value and risk study types

We saw earlier in this section that there is no one single type of value or risk study that suits all stages of a project. Not only will the objective of the study change through the life of the project but so will the information on which it is based and the participants in the study. This section outlines some of the key study types through the life of a construction project. The recommendations at the earliest stages will not necessarily include a construction solution. The study types are summarised in Table 4.4. Details of each study type are contained in Chapter 7.

**Table 4.4** Study types at different projects

| Project stage | Study type | Focus of value study | Focus of risk study |
|---|---|---|---|
| Inception | V-RM0 | *Need verification*<br>The trigger for a project is the realisation of the need for beneficial change. The first task is to assess whether a building project is the most appropriate way to deliver the required benefit.<br>The study involves a consideration of all the options that are available and testing these against the overall business strategy to verify that the project selected is the best way to deliver the benefit. | *Strategic outline case study*<br>At project inception, management needs to assess risks at a strategic level – risks that could affect the business in the long term, survival, continuity and growth.<br>At the same time, the risks of undertaking that project or the various other options need to be assessed.<br>The study will also develop strategies for managing the risks. |
| | | **The output of studies at this stage will provide recommendations to the management as to whether or not to proceed with the project.** | |
| Strategy and feasibility | V-RM1 | *Programme level definition*<br>If the decision at inception is to proceed by way of a number of interrelated projects (a programme) the first study will focus on making sure that all the projects conform to the overall programme objectives.<br>The study will identify the key value drivers necessary in order to deliver the programme objectives. They should only change if the programme objectives change. | *Programme level – preferred risk allocation risk allocation*<br>A risk study at this level will focus on setting criteria to ensure consistency of risk management across the programme. The study will identify appropriate risk categories, set limits for upward reporting of risk, parameters for quantifying risk and its preferred allocation. |
| | | **The output from a study at this stage will inform the programme brief with respect to clarity of objectives and allocation of risk.** | |

Continued

**Table 4.4** Continued

| Project stage | Study type | Focus of value study | Focus of risk study |
|---|---|---|---|
| Strategy and feasibility | V-RM1 | *Project level – definition*<br>If a project forms part of a wider programme, the value drivers provide the basis for building the project value profile. Where a project is stand alone, it will be necessary to develop the value profile independently.<br>The study will involve building the value profile to design considerations level. | *Project level – preferred risk allocation*<br>A risk study at this level will focus on setting criteria to ensure consistency of risk management across the project. The study will identify appropriate risk categories, set limits for upward reporting of risk, parameters for quantifying risk and its preferred allocation.<br>The study will be concerned not only with construction risks but all risks that could lead to project failure. |

**The output of studies at this stage will inform the project brief and lay the foundations for the management of risk.**

| Project stage | Study type | Focus of value study | Focus of risk study |
|---|---|---|---|
| Strategy and feasibility | V-RM2 | *Brief development study*<br>During this stage the design team will come up with a number of options to satisfy the project brief. Each of these will have its advantages and disadvantages. A study at this stage will help the team differentiate between the options and select the one that on balance has the potential to provide greatest value and least risk. | *Project review – procurement routes*<br>During the feasibility stage a series of risk reviews based on the parameters established at concept stage, will assist in selection of the most appropriate design option and individual project procurement strategies. |

**The output from a study at this stage will inform the design brief.**

| Project stage | Study type | Focus of value study | Focus of risk study |
|---|---|---|---|
| Pre-construction | V-RM3 | *Value Engineering Study*<br>During the design development the team may undertake a number of value engineering studies to optimise the chosen design solutions in terms of time, cost and quality. The output from each | *Design Stage Review*<br>Once the preferred design has been developed to a sufficient degree, successive risk reviews will monitor the effectiveness of the risk management to reduce exposure to risk to acceptable levels. |

**Table 4.4** Continued

of the value engineering studies will be a set of proposals for improvement to the design. These will be developed and validated by members of the design team and decisions made whether to incorporate them in the project or not.

**The output of studies at this stage will lead to the development of a design giving best value and least risk.**

| | | |
|---|---|---|
| Pre-construction or construction | V-RM4 | |

*Design and cost review*
Despite the project team's best endeavours, estimated costs for a fully developed design may exceed the budget. This study explores each of the functional areas of the building from the point of view of reducing cost and improving buildability. It is unwise to apply this technique in circumstances other than those where the design team fully understands the functionality of the different parts of the project.

*Construction review*
By the construction stage, much of the responsibility for managing risk will be passed to the contractor. He will need to manage these using the type of study outlined in V-RM3 above. There will however be a number of residual risks that concern the client body. These will include the potential for consequential risks from the construction project as well as those residual risks which were allocated to the client in the early stages of the project.

**The output of studies at this stage will reduce costs to acceptable levels and ensure that risks retained by the owner are managed effectively. Studies undertaken by the contractor will protect his profit margins and ensure that the finished building meets the owner's and operator's expectations.**

| | | |
|---|---|---|
| Handover/client occupation | V-RM5 | |

*Post-occupation reviews*
Once a building is commissioned and in use the team should conduct a post-occupation review to assess the effectiveness of the value management plan and the degree to which expectations have been met.
Later operational reviews may identify where improvements can be made.

*Post-occupation reviews*
Reviews, at this stage should assess the effectiveness of the risk management so that the team can learn lessons for future projects. They should also review the risks to which the client and users remains exposed and confirm plans for managing these.

**Studies at this stage will provide the basis for feedback for continuous improvement. They may identify the need for future projects to improve performance.**

All the above studies can be used very effectively to brief members of the supply chain who may not have been involved earlier in the development of the project. In traditional methods of procurement, this is particularly relevant later on in the design stage or during construction.

In all studies, if the previous stages have not already been undertaken, the study must establish those parameters, such as value and risk profiles, which would have been set up during the earlier studies. If project objectives or other factors change which affect these parameters, the earlier stages should be revisited to establish what impact the changes have on the project (see Figure 4.5).

## *Programmes of studies*

No study should be considered in isolation. Value and risk management should be viewed as on-going cycles of activity throughout the duration of a project (from inception to delivery of benefits). Study leaders should think in terms of programmes of studies, the output from each providing the input to a chain of activities which leads into the next (see Figure 4.2). Value and risk management should be part of 'business as usual'. This is not to say that studies can be informal. They should be formal, structured events. They should not be regarded as interrupting the process of project delivery, rather as assisting it.

## 4.9 Critical success factors

For a value or risk management programme to be successful certain key principles should be in place. These may be termed the critical success factors (CSFs) for the programme. All programmes are unique in their details. The following provide a useful basis to compiling the CSFs that apply specifically to your programme.

1.  Clearly identified and visible senior management support for the programme.

2.  Explicit policies which are clearly communicated to all.

3.  The adoption of a transparent and repeatable framework of activities.

4.  The existence (or creation, if necessary) of a culture that supports and understands the concepts of maximising value and controlling risk.

5.  Fully embedded management processes which are consistently and rigorously applied and are clearly linked to the achievement of objectives.

6.  Implementation of effective plans and regular reviews to ensure that the benefits of the processes are realised and lessons are learned for future programmes.

## 4.10 A framework for introducing value and risk management into an organisation

If an organisation is to actively manage value and risk it must ensure that the CSFs described above are embedded. This may be achieved by putting in place a framework of activities similar to that shown in Figure 4.7. The framework links value and risk management activities to an organisation's strategic goals. This ensures that programmes and studies are aligned with the

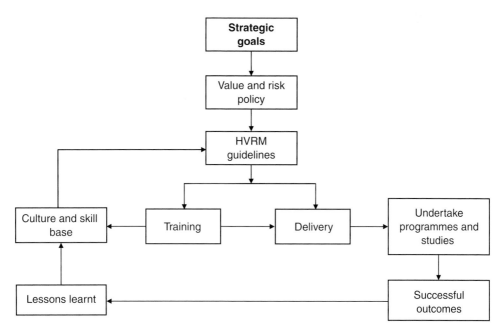

**Figure 4.7** The value and risk management framework

delivery of the goals. This section describes how to implement this framework.

## *Policy*

The policy should set out, among other things why the organisation supports the use of value and risk management and what benefits it expects to generate. These could, for example, be related to some economic trends, they have observed, be part of a broader package to enhance competitiveness or simply be a desire to follow best practice as promoted by such organisations as OGC or contained in the Turnbull Report.

A global manufacturer of consumer products noticed that the return that it was getting from its considerable investments in new production facilities was decreasing and, more significantly, falling behind that of its competitors. As part of an overall plan to redress this problem, the company decided to implement a value management plan on all new investments over a defined capital cost limit. To do this they charged a small group in their capital projects division to invite external value management practitioners to do three things:

- Develop and deliver training for senior managers and potential internal practitioners.

- Deliver value management services on new investment projects.

- Provide mentoring for the newly trained value management practitioners.

The results of the programme delivered capital and operating cost reductions averaging about 10% and made a valuable contribution towards achieving their overall objective of improving return on capital invested.

The policy should provide clear guidance on when issues or risks should be escalated upwards through the organisation, in case events occurring at lower management levels have consequences that might impact on activities at higher levels.

The policy should state, in broad terms, to which areas of the business the methods should be applied and provide guidance on the scale of that application.

It should state whether the organisation intends to generate its own internal delivery capability, rely on buying in the expertise when needed or a mixture of the two.

It should set out a timescale within which they expect to embed the practice of value and risk management into the organisation's culture.

## Champion

It is normal at this stage to appoint an individual to implement the policy. We shall designate this individual as the head of value and risk management (HVRM). Senior management should form a small steering group, which is representative of the various parts of the business, to whom the HVRM will report and discuss progress. Members of the steering group should report to the organisation's board.

The HVRM will have a sound understanding of value and risk management methods. He should draw up a plan setting out how he will implement the policy which the management has agreed. This will normally comprise two strands: training and service delivery.

## Guidelines

The HVRM will set out guidelines describing the types of study that should be conducted at strategic, programme, project and operational levels. These should be broadly consistent with the study types described elsewhere in this chapter. The guidelines should outline the process to follow, list suitable techniques, provide guidance on who should be involved and the level of competence and experience of the study leader who will lead the processes. The guidelines should provide the basis for delivering repeatable processes but not be so prescriptive as to stifle individual interpretation and innovation.

This is similar to a good cookery book. Those that give precise recipes in which exact quantities are given for all ingredients are notoriously difficult to follow. They do not encourage the development of intuitive cooking expertise. Those which describe the quantities in approximate terms (a handful of this, a tea spoon of that) and suggest alternative ingredients if you do not have the right one in the larder, are far less prescriptive and encourage innovation within the 'framework' of the recipe.

## *Training*

The first strand of the plan, training, will be shaped by the policy regarding the use of internal or external expertise. If the organisation only plans to use external expertise (e.g. because it is likely to be only an occasional user of the methods), the training programme will be focused on building up an awareness of the methods and their benefits at all levels of management. This is essential to gain support for the methods throughout the organisation and to build up a collaborative culture.

If it is intended to build up internal delivery expertise it will be necessary to train up or employ internal study leaders to competent delivery skills levels. Ideally, to ensure that the training is effective, it should be accredited by a competent organisation and lead to a professional qualification. Both the awareness and the practitioner training courses should align with the organisation's values and its approach to doing business.

The author's firm, Davis Langdon, successfully implemented a programme to develop value and risk management competence within its standard service offerings. It trained study leaders under accredited training programmes in most offices and regions across the United Kingdom and overseas. The programme also built upon and embedded a culture of awareness of value and risk across the firm. Clear and visible senior management support was, and remains, present at all times, culminating in a change in the Firm's Mission Statement to 'To add value and reduce risk for all clients'.

## *Delivery*

The second strand of the plan should be the delivery of the services themselves within the designated projects using the techniques described elsewhere in this book. The study leaders should gather feedback from all studies, in addition to the formal reports, so that they may build up an information base and learn from experience. The lessons learnt should be incorporated into the training programme and shared between the study leaders to ensure that the quality of service is continuously improved.

## 4.11 Embedding value and risk management in an organisation

Embedding a robust culture and the necessary skill for delivery of a successful value and risk management programme within an organisation requires a clear plan. A tried and tested plan was discussed under Section 4.9. It should be led by a suitably qualified individual, the HVRM.

## *Spectrum of acceptance*

Initially it may be difficult to persuade some people in the organisation to apply the value and risk management methods rigorously to their projects. Others, however, will leap at the chance, recognising the potential benefits and what these could do for their part of the business. The latter are known as innovators. The former may be non-adopters. In any organisation there is likely to be a spectrum of people between these two extremes as indicated in Table 4.5.

The HVRM needs to recognise differing appetites for adopting the new techniques and refine his plan accordingly. The innovators

**Table 4.5** Willingness to adopt change

| Characteristic | Behaviour |
|---|---|
| Innovators | Take up ideas without external push – need to be brave |
| Early adopters | Recognise the need but require guidance |
| Late adopters | Do not recognise the need but will follow once convinced |
| Non-adopters | Resistant to change |

and early adopters are unlikely to need much encouragement and, once they see the benefits flowing will become staunch supporters of the HVRM's efforts. By contrast, the late adopters will require significantly greater effort to cause them to break old habits and change to the new ways of working. With these people it is particularly important that their first experiences must be successful. This will require careful selection of the project in which they are involved and deployment of the most skilled and experienced study leaders available. Once later adopters have tasted success they are likely to take up the cause with enthusiasm and may become the most vocal ambassadors.

Non-adopters are resistant to change and will be the most difficult people to persuade. The HVRM needs to constantly make these people aware of the successes of others and show how these techniques could reap similar rewards for them. However, success is by no means certain.

## *Monitoring performance*

As the HVRM and the service providers gain experience, they will wish to modify their techniques to suit the circumstances. This type of constructive evolution should be encouraged provided it is not used as an excuse to short-circuit the rigour of the processes described elsewhere. Such temptations should be resisted since they are likely to lead to disappointing outcomes and a corresponding reduction in support for the processes. By contrast, the mark of the 'seasoned professional' study leader is the ability to flex and develop the processes to retain and enhance their rigour. Application then remains fresh and effective, avoiding the staleness that can develop without such innovation. It is clear from these comments that the HVRM should write the organisation's procedures for value and risk management in such a way that such constructive innovation is encouraged. Regrettably some really ground-breaking initiatives in large successful organisations have been stifled by the impositions of inflexible procedures and, in due course, abandoned.

A major infrastructure provider adopted a policy to undertake value engineering on every project exceeding £5 million in capital cost. The policy was mandatory. A team of internal facilitators was trained to deliver the service. Initially, it was very successful in reducing costs. However, as time passed, it became more and more of a 'tick box' exercise. The rigour and enthusiasm dwindled. The programme became less and less effective and was eventually stopped.

The successful innovations need to be shared with other study participants so that all can benefit from each other's experience. Similarly, innovations that were not so successful should be examined and shared, either to transform them into successes or to avoid others making the same mistakes.

## *Mentoring*

Obtaining a study leader qualification is only the beginning of progress to becoming a seasoned professional. Regular mentoring from seasoned professional will help him gain the necessary experience and depth. The HVRM is ideally suited to this role since mentoring can be a two-way experience. In return for giving his advice as to how the novice might handle different situations, the mentor will receive feedback on what did and did not work well. This way both parties will help to improve service delivery.

One of the most fruitful sources of innovation is to learn what successful practitioners do in other sectors of the economy. Often, with little adaptation, these techniques can be applied to building projects with great success. Examples of this are given in Chapter 6 and should assist readers to broaden their horizons in value and risk management practice.

An example of learning from other sectors is the use of failure modes and effects analysis (FMEA – see Chapter 10) to identify potential risks. The technique was originally developed to identify potential causes of failure in mechanical devices. However, the technique is easily adapted to identify risk in a building project.

## 4.12 Drawing on experience

The value and risk management culture will evolve with experience and maturity in the use of the processes. This will be assisted by building up an information base of things that have been effective in enhancing value and controlling risk in the past. Such information bases are notoriously difficult to build in a form that is easily accessible and understandable to those who were not involved. This is because they comprise two types of information – tacit and explicit. The explicit information is relatively easy to capture in a database, which may be searched with key words. This does, however, only give half the story. The tacit information is the first-hand experience that cannot easily be put into words. Normally, this information needs to be shared either through experience or in one to one dialogue between people who have had similar experiences.

> An example of the use of tacit information is understanding why the use of a workshop technique can be wonderfully successful with one group of people but fall flat with another. Another example is the difficulty one might experience in defining the critical point in a cooking recipe at which to remove a pan from the heat to prevent curdling the ingredients.

## Use of case studies and anecdotes

One way of capturing at least part of the tacit information is through the use of examples and anecdotes. These can capture the tone of the experience and, hopefully, enable the reader to translate the situations into his own circumstances. Another is to include in the information base, the contact details of those with first-hand experience of the subject in question so that the enquirer can discuss with them.

## Prompt lists

In Chapters 10 and 11 we discuss the use of value and risk prompt lists when generating ideas to enhance value or identifying risks. These comprise explicit information. It is good practice to constantly refresh prompt lists following successful risk and value studies since this is a way of building up and communicating experience.

## Seminars

One way to build on individual experiences in an organisation with a number of risk and value management practitioners is to hold regular gatherings where people can share their experiences. A useful format is where one practitioner presents a case study for others to discuss and question. It is good practice to compile a record of the case studies and discussions and add these to the information base. This enables any study leaders or other interested parties who were not present at the meeting to benefit from it.

## *Process maturity models*

Earlier in this chapter we related the extent to which organisations can add value and control risk to the maturity of those processes within the organisation.

To assess the maturity of processes within the organisation a number of maturity models have been developed. One of the best known of these is based upon work done at the Carnegie Mellon Institute. The APM Group Ltd has used this as the basis for developing the OGC Project Management Maturity Model, which is described in Chapter 7. This model measures maturity of processes on five levels (see Table 4.6). In practice few organisations achieve level 5. Figure 4.8 illustrates a similar approach to assess the maturity of competence in value and risk management.

**Table 4.6** Maturity levels in a Project Management Maturity Model

| Level | Processes | Description |
|---|---|---|
| 1 | Initial | Can the organisation recognise projects and run them differently to its ongoing business? |
| 2 | Repeatable | Does the organisation ensure that each project is run with its own processes and procedures to a minimum specified standard? |
| 3 | Defined | Does the organisation have its own centrally controlled project processes and can individual projects flex within these procedures to suit the particular project? |
| 4 | Managed | Does the organisation obtain and retain specific measurements on its project performance and run a quality management organisation? |
| 5 | Optimised | Does the organisation run continuous process improvement with proactive problem and technology management? |

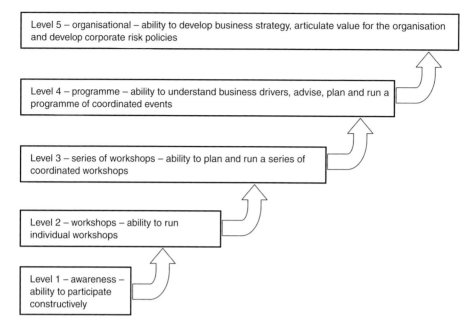

**Figure 4.8** Example of value and risk management process maturity levels

## 4.13 Selecting the appropriate level of activity

Although risk and value management will be most effective if planned throughout the life of the project, not all projects warrant the same level of activity. It is advisable to classify a project against appropriate criteria to assess the level of activity that is warranted. There are several tools for doing this. The cabinet office have developed The Project Profile Model, available at http://www.ogc.gov.uk, this enables users to identify problem areas and criticality to the business in terms of the scale of project against central criteria.

The following, very simple, selection tool allows the project manager to assess the project against four categories:

■   Size (expressed in monetary terms)

■   Complexity

■   Novelty or level of innovation

■   Strategic importance.

It is very quick and easy to use. The project manager assesses the project against each of these categories on a score of 1–5 (the range

1–5 being ascertained according to the typical work undertaken by the organisation). The aggregate score results in the project being classified minor (less than 10, medium 11–15, major >15). The range and level of value and risk management activities for each of these classes of projects should be set out in the organisation's value and risk management handbooks, compiled by the HVRM. Clearly, this tool is simply for guidance for the project manager; it is not designed to replace his experience and judgement. We give an example in Figure 4.9:

The project manager scores each category from one to five and adds the totals together to derive a project score. He then compares the score to the guidelines in the last box to select the scale of the project.

| Size | | *Complexity* | | Novelty | | Strategic importance | | Score (out of 20) |
|------|---|------------|---|---------|---|---------------------|---|-------------------|
| | + | | + | | + | | = | |
| 1. Up to £1m | | 1. Simple | | 1. Routine | | 1. None | | Guidelines: |
| 2. £1–£2m | | 2. Average | | 2. Common | | 2. Low | | |
| 3. £2–£4m | | 3. >Average | | 3. Unusual | | 3. Average | | >15 = Major |
| 4. £4–10m | | 4. Complex | | 4. Rare | | 4. High | | 7–15 = Medium |
| 5. Over £10m | | 5. Very complex | | 5. One-off | | 5. Essential | | <7 = Minor |

**Figure 4.9** Level of value or risk management activity selection tool

# **5** People

**Summary**

- This chapter explores how value and risk management interface with, and improve interactions between, the many stakeholders involved in constructing projects.

- The first sections look at the benefits of people working together in place of confrontation and how the study leader can overcome difficulties presented by different types of people.

- It touches on the role of stakeholder management and how the study leader should take these into account when planning the study.

- There is a section referring to team building and a description of the main features of partnering.

- It moves on to review the supply chain and the roles that each member may play in the processes.

- It looks at how to select an effective study team and who should be involved.

- It then looks at the role of the study leader, the key attributes that make a good one and the styles of leadership he should adopt.

- There is a section that stresses the importance of good communications both within the project team and with external stakeholders.

- Developing a value and risk culture that is appropriate to the organisation is vital if value is to be maximised and risks minimised. The section explores the characteristics of organisations with different risk appetites and how these impact on value creation.

■ The final section discusses the importance of good communications and use of appropriate language.

## 5.1 The people and their roles

Construction projects involve many stakeholders. Some of these will be directly involved with the project, others may have a peripheral involvement, others, the external stakeholders, may have no role in it at all, nevertheless are interested in its outcome. All need to play their part to assure the success of the project. This chapter looks at who these stakeholders are and to what extent they may be involved in a risk and value management programme.

For the project to succeed, those involved in the project and the external stakeholders who have an interest in its outcome must work constructively towards common goals. If they cannot work effectively and harmoniously towards agreed objectives there is likely to be confrontation and conflict. This, in itself, represents a risk to the successful outcome of the project and, hence, detracts from achieving maximum value (Figure 5.1).

The nature of the construction industry is such that most projects will involve new groupings of organisations and people. In the early stages the key players in the project delivery team may only recently have been appointed. They may or may not have worked together before. Regardless, everyone (including the owner) is learning about the new project. A value and risk management study at this stage (see Chapter 4) is a useful way to get the project off to a flying start.

## 5.2 Collaboration versus confrontation

The construction industry has never been renowned for its responsive and cooperative nature. Horror stories of clients and individuals having bad experiences with their professional advisors

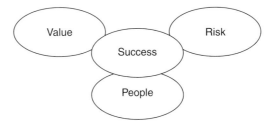

**Figure 5.1** Combining value, risk and people is the key to success

and builders abound. There have been numerous attempts to improve this situation but none had had much impact until the 1990s. In 1994, Sir Michael Latham was commissioned by the Department of Environment to address the situation. His first interim report identified two root causes for the problem – trust and money – too little trust between all the parties in the highly fragmented industry and not enough money to sustain a healthy industry. Latham's final report, Trusting the Team, proposed a series of recommendations. These included challenges throughout the supply chain, from clients all the way through to the tradesmen on site as well as legislation to outlaw certain unfair practices such as 'pay when paid'. A central message throughout was that of teamwork. All members of the supply chain should pull together as a team.

The momentum of improvement was increased when the Construction Task Force, headed by Sir John Egan, published its influential report 'Re-thinking Construction'. This set some stretching targets for the industry to achieve. Momentum quickly gathered pace as numerous initiatives promoted the new atmosphere in the industry. The Strategic Forum's 2002 report, Accelerating Change, sought to build on the successes to date and, among other things, set a target for 20% of construction projects by value to be undertaken by integrated teams and supply chains.

The entire modernising movement has been underpinned by a sense of responsibility to collaborate, rather than confront. The results have been a significant improvement in performance. Success is not universal. There are still those who exploit their suppliers while hiding behind the messages of the new era.

## Focus on outcomes

In the new environment, value and risk management, with their strong emphasis on collaboration, teamwork and trust, have not only thrived but have also brought a new dimension to the quest for improvement. Most of the targets set in Re-thinking Construction and Accelerating Change relate to improving the building *processes*. Value and risk management focus on *outcomes*. A successful outcome is the goal of all projects.

## Alignment with public sector initiatives

This focus on outcomes aligns with the clear commitment from central government, including HM Treasury, to promote better design across all public sector construction projects and the specific objective of optimising whole-life values, through the Prime Minister's Better Public Building Initiative. Details can be found at www.betterpublicbuildings.gov.uk. Further support for encouraging good design comes from the highly influential Commission for Architecture in the Built Environment (CABE) who have supported the development of the Design Quality Indicator, published by the Construction Industry Council, which went on line in October 2003 at www.dqi.org.uk. The Office for Government Commerce (OGC) has updated previous guidelines for value management in their Achieving Excellence in Construction programme which includes guides for best practice. Achieving Excellence Guideline No. 4 covers value and risk management. For further information on the OGC's programme, visit www.ogc.gov.uk.

## 5.3  Stakeholder analysis and management

Projects do not exist in isolation. All are subject to influences from outside the project team. The project exists within an environment, populated by the stakeholders introduced at the beginning of this chapter. Stakeholders' differing expectations can pose significant risks to a project. It is unlikely that the requirements of all stakeholders will coincide. Each will seek to influence the project in order to meet their own ends. Pressures from stakeholders can generate change, increasing costs and delay.

### *General principles*

Stakeholder influence is felt most keenly in the early stages of the project, when it is flexible at this stage and susceptible to change. It may drop off markedly when construction starts due to the perception that the project is a *fait accompli*.

On a sizeable project it is common to identify 40–50 stakeholder groups all with different involvement, requirements, levels of power to influence the project and levels of interest in doing so. This can be a very complex situation to manage. One technique to simplify the task is though stakeholder conferencing.

### *Stakeholder analysis*

To manage stakeholders it is necessary to understand whether they are supportive or not and how much power and influence they wield. The details of this technique are described in more detail in

Chapter 9. The broad principles, however, are to focus most effort on those of high power and high interest in the project. If these stakeholders support the project, ensure that they remain supportive. If they are against, much work may be needed either to gain their support or at least minimise their adverse influence. If they are of little power and influence, whether they are for or against the project, they should be kept positively informed with the minimum of effort.

The study leader should be familiar with the principles of stakeholder management since it is likely to form a crucial part of his study preparations.

## 5.4 Team building

Many people place very high emphasis on team building as one of the study leader's skills. In practice, taking into account the limited time that the study leader is likely to spend with the project team, he can have only a small influence on long-term team building. That said, however, any workshop or other team-based activity which the study leader may run within a value or risk management programme is likely to contribute positively to the overall building of the project team. The study leader's skills in resolving difficulties and brokering consensus will be particularly useful.

One of the crucial things to establish during any briefing session before a value or risk management exercise is the nature of the relationships within the team, who gets on with whom, who does not and so on. Such knowledge can play a key part in selecting the membership of workshop teams or working groups and the smooth running of a study.

Earlier in this chapter, we described how the study leader will help the team to work together. We have also explored how he will encourage active management of all stakeholders within the project and externally. Collaboration, rather than confrontation is a key principle of both value and risk management.

## 5.5 Partnering

One aspect of collaborative working is known as partnering. Many study leaders are skilled in partnering facilitation. Because many of the aspects of a partnering workshop can be of great assistance in getting a team to work well together, even if they are not formally partnering, it is worth including a few words on partnering here.

The concept of partnering is encouraged within the Re-thinking Construction agenda because it has the potential to cut out all the wasted effort of traditional low trust, confrontational relationships. Endemic fragmentation in the construction industry was identified by Sir Michael Latham as a key source of waste.

The key to partnering, whether formal or informal, is that all stake-holders throughout the supply chain (the partners) evolve a culture that encourages them to work constructively together towards delivering a common objective. This culture requires that they build trust in each other and, as a result, communicate openly. Open communication includes sharing financial and other information, normally kept hidden, with other partners trusting that you will not suffer competitive disadvantage as a result.

## Value for money

All partners learn from each other and build more productive working relationships. This will lead to more *efficient* operation. Elimination of wasted effort will lead to more *economical* delivery. Better understanding between all will lead to more *effective* solutions. These conditions thus fulfil the requirements of the National Audit Office's requirements for value for money: efficiency, economy and effectiveness.

Where partners are involved in several projects, investment in common systems, training and research and development can lead to continuous improvement and innovation that is measurable and against which targets may be set.

## Benefit trading

Trust between large and small organisations throughout the supply chain can open the door to benefit trading through which all will profit. For example, a large organisation may be able to obtain large discounts on materials that are used in the manufacture of goods by one of its suppliers. By providing these discounted materials to the supplier, it can receive the goods at reduced cost without reducing the supplier's margins.

## Innovation

A sharing of intellectual capital will contribute to the generation of innovative solutions to problems, one of the key objectives of value management. In traditional relationships, a significant amount of time is spent developing competitive bids for work. The concept of framework contracts for delivery of services within a single client body can reduce this. The concept of partnering, however, takes this one step further, introducing the concept of trust (often missing in framework arrangements).

## Risk and reward

The concept of sharing risk and reward is one of the key principles of partnering. In practice, this is probably the hardest part to achieve. In a true partnership, one person's problem is everyone's problem. All should try to help each other overcome each other's problems, even when they arise from another's errors. In this utopian scenario the time spent on monitoring, preparing and fighting claims reduces, thus freeing up time for constructive activities.

Resource planning is easier because the contractor and suppliers work to an optimum timetable rather than an ultra-safe timetable. This can lead to *quicker delivery*.

> The Eden Project was procured under the NEC form of contract which entitles the contractor to compensation events should certain risks occur. In this case the contractors, Sir Robert McAlpine and Sir Alfred's joint venture was incentivised to minimise the incidence of compensation events. This, combined with other active measures to manage risk, resulted in only half the risk allowance being spent.

## Contracts

Some suggest that contracts may no longer be needed in a true partnering culture. In practice, it is likely that contracts will always be needed to provide the underlying principles and ultimate sanctions against a 'partner' who chooses to defy the partnering concepts. The contract should, however, be of a type that permits a clear definition of an equitable sharing or risk and reward. The New Engineering Contract (NEC) and PPC2000 are contracts of this type.

Some of the traditional 'adversarial' contracts have published 'partnering' supplements. While these provide a step in the partnering direction, they are, in effect, sticking plaster remedies. They do not, for example, override the contractual requirements for parties to submit apparently confrontational documents within specified time limits or forfeit the right to compensation. They are therefore, ineffective.

## Partnering workshops

One of the most effective ways of introducing the partnering ethos into a project is by way of a series of partnering workshops. Value and risk management study leaders possess the skills to deliver these effectively.

The power of the team, working together collaboratively, over the individual is captured in the following quotation:

Alone we can do so little, together we can do so much. Helen Keller

More details of the techniques of running a partnering workshop are given in Chapter 9.

## 5.6 Communications

In just about every project review and in every 'Construction Process Improvement' workshop, communication is identified as a major issue (and, usually, not because it is good!). Good communication between all those involved on any construction project is vital. Earlier we identified the terms and acronyms used in different market sectors as roadblocks to good communication. Clearly, communicating in language that all parties can understand is crucial. One of the critical success factors in value and risk management is that those involved can learn from each other. Learning requires good communication and clear messages. Because value and risk management studies help all participants to communicate with each other they provide an excellent opportunity to learn.

Project communications fall into two essential categories, external, between the project team and the external stakeholders, and internal, between members of the project team.

## Internal communications

Communications start within the client body. Clarity in what the project must achieve is essential if there is to be any chance of communicating it to others.

Next, it is necessary to articulate these requirements in an unambiguous manner to the project team. This is particularly important where the client wishes to benefit from the experience and creative potential within his project team by specifying his requirements by outcomes rather than solutions. Prescriptive statements of Requirements may provide clear and unambiguous instructions to the project team but they allow little room for innovation. Performance-based specifications are better because they allow freedom to choose between several different solutions, each of which delivers the required performance. Best of all is a specification that describes the required outcomes with minimum restraints on solutions that can deliver these. Developing such a specification in a manner that allows freedom for innovation and, at the same time, is unambiguous provides a major challenge. One technique that is popular in France is function performance specification. This is described in Section 10.16. A simpler vehicle for articulating the client's requirements unambiguously and clearly is by the use of value profiling and the value cascade (see Sections 6.1 and 10.8).

## Communications plan

Good communications within the project are no less important. It is normal to include a communications plan within the project handbook. This should include a project directory with contact details and summary of roles of all members of the team, a description of what communications should pass between whom (nature of communication, method and numbers of copies of documents and so on), how decisions are made, how to escalate the level of decision-making if this is necessary and the problem resolution procedures. The plan will also set out the schedule of meetings and who should attend them. It will also describe the regular reporting structure, who will receive the reports and why.

The value and risk management process can play a vital role in providing a means of helping all team members learn from each other and resolve problems. In particular, they will throw light on requirements and other potential problems.

## External communications

Communicating effectively with the external project stakeholders is essential if the project team (including the client) is to gain and retain their support. Stakeholder management represents a very positive way of doing this. Again, the key to good communications lies in articulating a clear message and developing a good Communication Plan.

## 5.7 The supply chain

The construction supply chain (Figure 5.2) comprises more than the contractor, his trade contractors and product suppliers. It should embrace *all* stakeholders, starting with the end users. It is the end users of the building who should have the greatest influence on what is to be built since it is they who will occupy it in the long term. The building operator should also be involved, since it is he who will be responsible for maintaining and operating the facility. The owner has a most significant role to plan because he pay for its design and construction. The project team, comprising the client representative and advisors, the project manager, the lead designer (architect

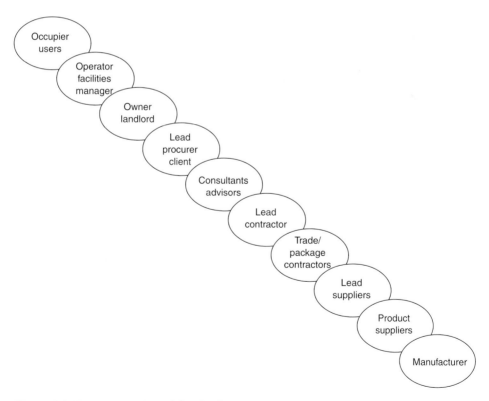

**Figure 5.2** The supply chain (after Be Benchmarking Report, June 2003)

or engineer), the supporting and specialist design disciplines all have crucial roles to play.

## The users, operators and owner – who leads?

In many projects, the client body is not a single entity with a single identifiable need. Large corporations and organisations have many different departments. Each of them will have different needs from a built facility. Balancing these needs to obtain the optimum facility for the business as a whole can be a time consuming and tricky process. It is, however, an essential part of defining the brief. The client body may include those who will pay for the project, those who will use the facility, their respective advisors and third parties with significant interests in the outcome of the project. The 'client' therefore is often no single body of interest but a multiplicity of complex, and often not complementary, interests. Then there are all the other supply chain members, illustrated in Figure 5.2 as well as the external stakeholders. A hospital provides a good example. Who is the real client? The primary healthcare trust? The NHS trust? The local community? The consultants? The nurses? The patients? For whom should value be maximised and risk minimised? Clearly, the interests of all of these groups must be taken into consideration and a balance struck.

## The occupier or users

The end users will comprise all those who will use the completed building. These may be office workers, factory staff, public servants, professionals, such as healthcare specialists, or members of the public. This group will enjoy the use of the building in the long term and it is vital that it works well for them. In some cases, for example, speculative developments, the end user will not be known. In such cases, their interests are represented by agents, who have the task of finding tenants for the owner or landlord.

## The operator

The next group comprises those who must run or operate the building. They will be responsible for ensuring that conditions within the building are right for the users and that the building is maintained in good order. These are the building operators or facilities managers.

## The owner and lead procurer

Central to the development is the owner who will commission the work. Without him there will be no project. It is vital that the requirements of the owner and those who fund the project (who will often be a separate organisation) are met. These stakeholders may or may not be the same as the organisation as that which procures the project. For example, a facility might be built for one of the divisions of a large company but procured by a separate property and construction division.

## The consultants and advisers

There will then be the team of consultants and advisers. On major projects there is usually a project manager who acts as the primary communications channel between the client body and the project delivery team. He is a key link in the communication chain and will play an essential part in any value and risk management programme.

The next group of stakeholders comprises the design team, usually led by the architect or engineer, and cost consultants. It is they who will translate the clients' needs into an optimal design solution. They may or may not be in place at the outset of the project (initiation) but need to be closely involved from the strategy and feasibility stages onwards. In conventionally procured projects the contractor is not engaged until the design is substantially complete. In modern procurement routes it is becoming more common to involve the contractor at an earlier stage, in a consultant capacity, to gain the benefit of his practical expertise during the design development. The end users of the completed project, together with specialist advisors, should be closely involved from the outset and throughout the design phase.

## The contractor and his supply chain

Once appointed, the whole of the contractor supply chain should be embraced within the value and risk management activities, which address construction (although they may not be involved in the client side programme). These will include the lead contractor, trade and package contractors, suppliers of materials and equipment, manufacturers and so on. This group supplies, installs and commissions the hardware that constitutes the finished building.

Typically, they control more that 70% of the costs of construction. For this reason, in Chapter 9, we discuss the concepts of target costing and working with the supply chain to optimise the value added within it. This approach is relatively unused in the construction industry. It holds great potential, provided the supply chain is genuinely involved in the quest for added value and reduced uncertainty, while protecting their own margins.

## *Others*

In addition to the supply chain, there are those who have no direct part in delivering the project but who have an interest in it. There may be numerous third parties who may have an influence on, or be affected by, the project. These range from the regulatory authorities, whose permission is required to build and operate the facility, to neighbours who live or trade alongside and members of the public who may take a close interest in the outcome. All of these should be consulted in the value and risk management programme to ensure their cooperation. Their involvement may give rise to unexpected benefits to the project delivery team.

A water company needed to lay a new main along a route that crossed many farms and other landowners' territory. They could have invoked their statutory right to force the landowners to give them easements along their preferred route for the pipeline. Instead they chose to consult with the landowners about their plans and agree a route that minimised disruption to them. In return the landowners suggested improved routing across their lands, avoiding many of the underground obstacles that would otherwise had to have been overcome. Although one landowner would not cooperate, forcing the water company to invoke their powers overall, the result was reduced costs for the water company and cooperation from most of the landowners. Surely this was a better outcome than the more common confrontational approach adopted in these situations.

Complex groups of stakeholders may reconcile their interests using stakeholder conferencing (introduced in Chapter 4) before any project is commissioned. Many projects will employ active stakeholder management described later in this chapter.

## 5.8 Selecting the study team

As a general principle, value and risk management studies should involve the full spread of stakeholder who can make genuine contributions towards achieving the study objectives. Stakeholder conferences may involve 50 or more people. More focused studies may be limited to much smaller numbers. The following provides guidance on selecting the optimal study team.

## *Selection criteria*

The ideal team should comprise the right mix of technical disciplines and a mix of complementary characters. Selection should be done in conjunction with the client body, and made on the following broad selection criteria:

■ Contribution to the study (reflecting technical knowledge and experience)

■ Attitude and character (how they will interact with each other)

■ Familiarity with the value or risk management process

■ Genuine stake in the outcome of the study.

The number of participants in a workshop can vary from a compact group of 6–8 people, to a large body of stakeholders numbering 50 or more. The same rules of selecting individuals should apply regardless of the size. The temptation to acquiesce to the inclusion of supernumeraries, who do not conform to the criteria listed above, should be resisted. The general rule should be to keep the workshop as compact as possible, consistent with fulfilling the above criteria.

The study leader should acquaint himself with the lines of responsibility in the contributing organisations to ensure that he speaks with the person with the authority to select the workshop and study team members.

To ensure that all aspects of the project are represented, the study team should include:

■ The commissioning organisation (i.e. the organisation that pays for the project)

■ Those who will use it

■ Those who will manage the delivery

■ Those who will design it

■ Specialist advisors (e.g. legal, marketing and land) and

■ Other departments or organisations who maybe directly influenced by the project.

The stakeholder consultation process referred to earlier should, in addition to the above, include external stakeholders whose influence or statutory powers could enable them to have profound influence on the project outcome. The identity of these should be ascertained in any exercise involving stakeholder management and may include:

■ Local statutory authorities

■ Residents associations

■ Local trade and business associations

■ Adjoining landowners and tenants.

## Personalities

Ideally, the study leader should select the workshop participants using objective character assessments. Unfortunately, this is not always realistic due to constraints on time and resource. He will usually have to work with the core members of the project team, regardless of their character profiles. If he does have the chance to select on the basis of character, two common methods of assessing character are the Belbin method and MBTI. The reader can learn more about the details of these techniques by reference to the sources given in Appendix A. Both methods give the study leader a good insight as to how people will interact during a workshop and provide an objective method of making a choice between two individuals of similar technical ability. Regardless of the method used (and intuition and recognition are probably the most common), teams should as far as possible be balanced in terms of individual characteristics. A majority of like-minded individuals can dampen the interactive dynamics on which a successful workshop depends.

## Authority

A workshop is an intense event. Provided preparation has been done adequately beforehand, it will require decisiveness and originality among the contributors if it is to be a success. For this reason, workshop participants should be empowered by their organisations to make representative statements and make decisions. Nothing is more frustrating than a member of the team who must refer every comment to a superior before responding to the workshop team.

This last requirement generally means that individuals will be fairly senior in their organisations. This is an underlying reason for insisting upon careful preparation before any workshop so that the workshop itself is fast moving, rigorous and decisive. This will keep the time and resource required for it in line with its importance.

## 5.9 The study leader

For a value and risk management programme to be successful the choice of the study leader is critical. His role is crucial but subservient to those who will deliver and use the facility. The study leader's primary role is to design the format and processes of the value and risk programme in the most effective way and then use all his skills and experience to guide the project team through the process to a successful outcome.

This book refers to the person who leads value and risk studies as the study leader. Another commonly used term is facilitator. The term study leader is preferred because value and risk studies include more activities than workshops alone. Workshops require facilitation skills. Preparation, reporting, implementation and review require leadership skills. The study leader should have both.

The good study leader should, first and foremost, be fluent in the delivery of value and risk management processes. He should have qualifications, issued by recognised professional institutions in both subjects. The seasoned study leader will have several years of experience in successful delivery of the services in sectors of the economy that are relevant to the subject under study.

It is not the study leader's role to provide technical advice directly, although he may provide advice by way of suggestions to the project team.

## *The attributes of the study leader*

Value and risk management processes are specialist skills and should be applied rigorously if the full benefits are to be achieved. To ensure this rigour, they should be run by an accomplished study leader, who is competent in the specialist skills of facilitation and who is fluent in the risk and value management processes themselves.

## Commanding respect

The study leader should command the respect and confidence of the people with whom he is working. This ability comes with experience and excellent communication skills with people from all levels. Communication skills include the ability to compile effective presentations, the ability to listen actively and summarise what he has heard.

## Positive attitude

The study leader should be an accomplished facilitator with a positive, 'can do' attitude, inquisitive, willing to listen, understand and give credit for the contributions of others. He should be a confident self-starter, with the ability to motivate others.

## Ability to adapt

The competent study leader, with experience, should be able to adapt the techniques to suit the circumstances and the subject under consideration, without losing the rigour and effectiveness of these techniques. This is very different from taking shortcuts to reduce the time taken or the cost of a study to suit the short-term aims of members of the project team. Such short-circuiting of established rigorous processes is a threat to the credibility and acceptance of value and risk management, and should be resisted at all costs.

Once a team has experienced bad or sloppy service, they will be reluctant to use it again (even properly) because it did not deliver to their expectations.

## The seasoned professional

The term seasoned professional is used to describe the study leader who has not simply gained qualifications in risk and value management but has a wealth of experience of use of the different techniques in a wide manner of circumstances such that he can make a sound judgement as to what is likely to deliver the best results in the circumstances. Seasoned professionalism is something that can only be gained through significant experience and is the goal to which all professional study leaders should aspire.

## Objectivity

A key attribute of the competent study leader is objectivity. While he may well provide advice based on his or her broad experience, he should never attempt to impose his views on the workshop team. To do so would be to take away from the technical team their primary role, which is to deal with the content of the study and to gain ownership of the study's recommendations. He should not side with any of the stakeholders who are represented in the study, but rather be even-handed with all. The study leader should remember that it is his primary role to manage the process that enables the team to evolve its own recommendations.

## Humility

The study leader should not seek the glory for conducting a successful study for himself. He should be generous and even-handed in acknowledging the contributions from all the members of the study team.

## 5.10 Styles of leadership

The accomplished study leader will need to change his style at different stages through a typical study. For example, during the strategic briefing meeting, a key step at the outset of all value or risk studies, the study leader is likely to adopt the style of a consultant. He will advise his client on the process that he is recommending, and will elicit information in a professional manner to inform the study. He may use audio/visual aids to assist in putting over his message and make it more understandable. Essentially, his role is to advise and educate those who are unfamiliar with the process and gather and process information to help it succeed.

During a workshop session in which the objective is to generate ideas to improve value, or identify risks that need managing, the study leader will adopt a pure facilitating role. He will encourage the team to make positive or constructive contributions but it is *their* contributions that should be recorded in their words. Here again the experienced study leader may make suggestions based on his experience but not include those suggestions in the output unless they are fully supported and owned by members of the team. Table 5.1 indicates the study leader's different styles at different stages throughout a typical study.

## 5.11 A study structure based on logic

The stages in a study, whether it may be risk or value, follow a common thought pattern. Initially, the study is all about consultation, expanding the quest for information that is relevant to the objectives

**Table 5.1** Study team leader's style during differing study activities

| Task | Consultant | Controller | Facilitator |
|---|---|---|---|
| Pre-workshop briefing | ✓ | | |
| Gathering and analysing information | | ✓ | ✓ |
| Arranging the workshop | | ✓ | |
| Workshop introductions | | ✓ | |
| Workshop team briefing (by project team) | | | ✓ |
| Function analysis | ✓ | | ✓ |
| Risk analysis | ✓ | | ✓ |
| Idea generation | | | ✓ |
| Idea evaluation | | | ✓ |
| Risk management strategies | ✓ | | ✓ |
| Idea development | ✓ | | ✓ |
| Risk management action selection | ✓ | | ✓ |
| Decision building | | ✓ | ✓ |
| Reporting | ✓ | | |
| Monitoring implementation | ✓ | | |

of the study. There then follows a summarising and ordering of that information to provide easily understood information as input to the workshop. At the beginning of the workshop the boundaries of thought will again be expanded during the briefing stages and discussion of the information, which has been tabled. There follows then another period of analysis, which brings focus to the study. When generating ideas to add value, or exploring management strategies to mitigate risk, the mood of the workshop again expands. It then contracts again as the team is asked to evaluate the various ideas that have been generated and assess which management actions are most appropriate. Finally, in analysing the output from the workshop and validating the proposals that have been made, the team expands its focus to explore all likely options before finally reverting to analytical mode to make its recommendations. It is this process of alternate expansion and contraction, or focusing of the thought processes within a workshop which is one of the keys to gaining a successful outcome.

The expansion, or creative thinking, described above involves the right-hand side of the brain. The focused or analytical thinking involves the left-hand side (Figure 5.3). Thus the logic of alternating the thought processes during a study is underpinned by psychology. The logical progression is one of the main reasons why taking short cuts reduces the effectiveness of the methods.

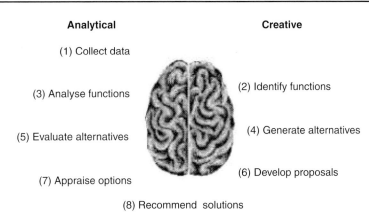

**Analytical**                                    **Creative**

(1) Collect data

(3) Analyse functions                         (2) Identify functions

(5) Evaluate alternatives                     (4) Generate alternatives

(7) Appraise options                          (6) Develop proposals

(8) Recommend  solutions

**Figure 5.3** Left and right brain thinking

## 5.12 Overcoming difficulties

One of the most difficult areas for the study leader to cope with is overcoming difficulties that may arise during a study. Difficulties can arise for many reasons. There may be a lack of belief in the process. There may be difficult people present. Members of the team may be working to different agendas which are not aligned to those of the study. If the project team, is not acting harmoniously as a team, there may be conflicts between individuals and a tendency toward point scoring. Encountering difficulties such as these can be very unnerving for the study leader and disruptive for the study. The difficulties may be compounded by the fact that he may not know many of the members of the team. He must deal with these situations to prevent the study being blown off course. Some reasons for conflict are summarised in the examples in Figure 5.4.

There are numerous textbooks on conflict resolution and ways of dealing with the above type of situation (see references in Appendix A), however, the general rules are as follows:

■ The study leader should separate emotions and people issues from the problem. He should discuss and consider the various aspects of the problem from both sides' perspectives.

■ He should listen for emotional responses, try to understand why those emotions are felt and get the person feeling them to express his feelings clearly. It is important to be even-handed when speaking to people within a group, and not favour one party over another.

---

**Misunderstanding, lack of effective communication**

'I cannot give my views or influence decisions in any way if I do not understand what's going on and the reasons for it.'

**Differing objectives**

'You may want to save money on installation, but I've got to run this b*****y thing!'

**Blaming others**

'It's all the project manager's fault.'

**Individual personalities**

'He's an arrogant know it all – I'll show him.'

**Attitudes**

'I'm not having that young Project Manager tell *me* what to do.'

**Expectations**

'I thought we would get more detail than that.'

**Insecurity, needing to make mark**

'I must show up the Project Manager so they think that I'm as good as he is.'

---

**Figure 5.4** Some reasons for conflict

- The study leader should focus on people's interests and not their positions. Positions are normally taken as a way of avoiding talking about the problem. He could make a list of each protagonist's interests (not their positions) on two separate sheets of paper. He should identify the common points of interest, not focus on the differences. It is likely that the points of common interests far outweigh the differences.

- He should ask the group to explore a variety of options so that each party gains something (the win–win solution). He should try to avoid any premature judgement or a single answer. There is usually more than one way to solve a problem. Encouraging each protagonist to consider the other person's problem as well as their own can often help them understand it better and allow them to accommodate at least part of the other person's requirements.

- When closing the discussion to reach an agreement, he should use objective criteria – making comparisons with, for example, market conditions, precedence and published data rather than parameters that are drawn directly from the problem. He should draw upon norms, benchmarks and standards. If one party's demands are seen to be way out of line with common practice, it is more likely that they may relax their demands.

Generally, the participants in a value or risk management study are selected on the basis of the technical or managerial expertise they

*The spectator* – does not easily participate.
Be patient, give role in warm up exercises, ask questions directly. Try to draw them into the discussions.

*The windbag* – tends to try to dominate discussions.
Limit time for discussion, target questions to others.

*The rambler* – loses track and uses far-fetched examples to make point.
Ask how to capture in one line on flip chart, move on during pause.

*The squatter* – will not change from initial position.
Discuss consensus building, enlist help of others, provide an exit route.

**Figure 5.5** Some difficult personalities

can contribute, rather than their skills in contributing to a collaborative effort. As a result of this, it is not uncommon that there will be some difficult personalities whom the Study Leader will need to persuade to cooperate constructively in the process. Some examples of these personalities are given in Figure 5.5.

One of the key tasks of the study leader is to get everyone in the workshop to make contributions appropriate to their knowledge and discipline within the project team. Sometimes, very knowledgeable and experienced people do not participate willingly, due to shyness or fear of being shown up.

> Better keep your mouth shut and be thought a fool than open it and remove all doubt. Denis Thatcher

## The spectator

When confronted with a team member whom the study leader believes has much to give but is reluctant to contribute, he should be patient and invite the person to share his experience with the team. Encourage him to show-off their expertise in non-threatening circumstances. The study leader should not show them up by asking such questions as: 'Well Fred, I haven't heard much from you, haven't you got anything to add on this situation?' This is unlikely to encourage them to emerge from their protective shell.

## The windbag

The opposite type of person illustrated in the previous situation likes the sound of his own voice and always tries to dominate proceedings. That person is often the expert on everything and stifles contributions from other team members. A technique that can be effective is to limit time for discussion; deliberately, direct questions at others to bring them into a conversation.

## The rambler

Very often technical experts do not make good orators. In describing what they want to share with the team they tend to get bogged down in detail and go off the point, using perhaps far-fetched examples to try and make a point, which no one understands. In this instance, the study leader might wish to ask for a shorthand version to summarise what the contributor is saying.

## The squatter

A frequent obstacle is the expert who has very fixed views and has no interest in changing from their initial position. Such a position can stifle innovation or prevent recognition of real risk to a project. The study leader should first try to enlist the help of others in exploring alternative solutions or ideas, and seek an exit route for the obdurate individual. He can then claim that he was right all along even though the team may wish to explore alternatives.

## 5.13 Using conflict to advantage

Not all conflict is counterproductive. One of the keys to a successful workshop is to select a multidisciplinary team with different backgrounds and attitudes. Apart from ensuring that all necessary technical expertise is present during the workshop, this is likely to introduce differing views and solutions to the problem being studied. One of the challenges for the study leader is to harness this 'conflict' constructively, leading to innovative thinking and improved outcomes. Handled correctly, conflict can also engender competition between different team members, helping them strive all the more towards their goals.

## Use of independent team members

In the United Kingdom it is generally accepted that value and risk studies are conducted in collaboration with the incumbent project delivery teams. This is to ensure that they own the resulting actions. Introduction of one or more independent experts can provide very helpful and constructive challenge to the incumbent team provided it is done sensitively so as not to trigger confrontation.

## 5.14 Gaining consensus

One of the most crucial tasks of the study leader is to bring a study to a close rather than leaving lots of loose ends of unfinished business. This means helping the team to build decisions and make recommendations. We have seen earlier that one of the benefits of the workshop approach is that it enables the team to build consensus decisions, taking the better parts from a number of views and constructing a solution which is better than that proposed by any individuals within the workshop. Consensus is vital, however, it is not the same as unanimous agreement.

The first step in achieving consensus is to ensure that each participant's views has been heard fairly and taken into account. There are a number of techniques for building consensus. Each of these helps take the team to the position when every member can give a clear yes to the following questions:

- Will you agree to the next step or question?

- Can you live with this position?

- Are you comfortable with the course of this action?

- Can you now support this alternative?

When this position is reached the study leader should record the fact and move on to the next issue.

### *Consensus versus voting*

Consensus is better than voting (which some people used to gain decision in a group of mixed interests), because it does not involve winners or losers. If, in a team of 12 people, 10 prefer one course of action and 2 people vote against it, there will be 2 dissatisfied members of the team. Because they are unhappy they may spend much of their time trying to undermine the decision. This causes dissent within a team and is not helpful. It is far better to use one of the techniques described in Chapter 9 to achieve consensus where the same two members of the team can answer yes to the questions posed above.

### *Consensus versus compromise*

Consensus is not the same as compromise. The art of building a consensus requires identifying the best features among the issues under discussion and putting these together into a proposal that all team members can support. In principle, it is like finding the highest common factor in mathematics. A compromise, on the other hand, involves reducing the proposed change to a level that it is acceptable to all parties. It is akin to finding the lowest common denominator.

## 5.15 Developing value and risk culture

A key factor affecting an organisation's approach to both value and risk management will be its culture. In an established organisation it is likely that a distinctive culture will have built up, either through formalised training and other culture building events, or simply through the way in which the management and staff interact and go about their business. Usually the organisation's culture will closely reflect the values and ethics held by senior management. Ingrained culture can be difficult to change. The process of change must be carefully managed and is likely to take a long time.

The value management standard BS EN 12973 : 2000 emphasises the need for a healthy value management culture if value management programmes are to succeed. It requires that each organisation will have its own unique:

- Visible commitment by the highest levels of management to considerations based on value.

- Positive changes in attitude and working practices to minimise waste, a commitment to improving value, and certainty.

- Understanding throughout the organisation of what represents value and how to achieve it.

- Approach to responding to the needs of internal and external customers.

- Desire for continuous improvement.

The above attributes, with a little imagination, apply equally to the make up of a healthy risk management culture. One thing is certain, if those in the organisation have differing views on value and risk there will be little consistency in the application of the processes and, as a result, a wide variation in the benefits achieved.

The successful organisation will develop its value and risk management processes to match the extent to which it is prepared to

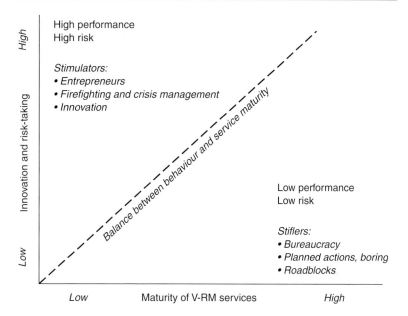

**Figure 5.6** Value and risk management maturity

innovate and take risks. Value management provides the means to define value and then encourage innovation to deliver it. Risk management allows successful and controlled risk taking, thus enabling the implementation of innovative solutions. A balance between maturity of process and the extent to which the organisation innovates and takes risks is represented graphically by the diagonal line in Figure 5.6.

An organisation which has few repeatable processes in place may wish to be at the forefront of innovation and risk taking but is likely to lack the tools to innovate successfully and may not be able to control the risks it is taking. If it succeeds in avoiding the risks (largely by intuition) it may be very successful. If not, the results can be disastrous. It is not held back by 'unnecessary' procedures or rules and can be swift to exploit opportunities in the market place. This is an entrepreneurial organisation – high gain, high risk and, unfortunately, high failure rate.

This type of organisation was typified by the Dotcom pioneers in the late 1900s. Many were highly entrepreneurial, had few robust management systems in place. They were very innovative in their offerings. Potential performance seemed to have no limits and investors piled in. Then things started to go wrong and most crashed into oblivion.

By contrast, the organisation that has many procedures and processes in place but is conservative and takes few risks is likely to be a very safe but relatively dull environment in which to work. The weight of procedures and red tape discourages innovation and builds a risk adverse culture.

> This culture is widespread in central and local government as well as large mature organisations, such as insurance companies. These organisations need to be seen as sticklers for detail and conformity with the law. They have extensive bureaucracy and sets of rules, which all employees must rigorously follow. The result is that things tend to move slowly and opportunities for innovation are discouraged. However, corporate failure is rare.

The balanced organisation will develop its processes to match the growth in its innovation and risk taking. This organisation is likely to grow successfully, taking controllable risks and reaping the rewards by adding maximum value.

## Risk and opportunity

The Turnbull Report recognises that if organisations are to prosper they must take risks. One of the main aims of the report is to ensure that the organisations have in place the mechanism and processes to keep those risks under control. Profits are the reward for successful risk taking.

The aim of risk management is, therefore, not to eliminate risk but to control it. Successful risk management reduces the uncertainty in achieving a successful outcome to acceptable and manageable levels.

## 5.16 Language

Every section of our community develops its own way of expressing itself. It develops its own language, or jargon, in which some words acquire specific meanings that may not be the same as the meanings those same words have in every day English. For example, take the word programme. In the building industry (in the United Kingdom) the programme refers to the timetable within which certain activities are expected to take place. To the layman it is more likely to be associated with television. In the United States the same word (spelt slightly differently) means the scope of works within the project. There is plenty of opportunity for misunderstanding.

A related cause for confusion is the use of acronyms. There is an increasing trend in all sectors to use acronyms as a form of shorthand to describe attributes, processes or events. The trend seems to be particularly strong in public services and the military. However, the Information Communications and Technology sector is not far behind. Indeed the whole industry is commonly described by the acronym ICT.

Then there are the technical terms which, to the uninformed, are frequently referred to as jargon. Value and risk management have developed their own jargon – function cost analysis, Monte Carlo simulation and so on.

It is essential that those offering to deliver services to organisations outside their own sector understand their client's language. Conversely, those in the building industry tend to forget that the majority of those people who want buildings come from other sectors. We tend to use language in reports and other documents that is hard for 'lay' customers to understand, or present figures in ways that do not relate to the customers accounting categories. This is sometimes done for exactly those reasons – to confuse the picture and thereby escape blame. In the post 'Re-thinking Construction' era of open communications and trust we should be using language and presenting information in ways that all can understand.

The construction of a value cascade, linking project objectives to elements within the building can help this process.

## Understanding the customer's needs

Chapter 4 emphasised the importance of understanding the customer. Use of common language is one facet of this task.

Within value management, the main tools for building and understanding a client's values are the value propositions, the value cascade and value profiling (see Sections 6.2 and 10.8). These models help the team to understand and articulate the client's strategic imperatives so that they can respond through the design and delivery of the building.

On the risk management side, it is important to build an understanding of the client's appetite for risk (see Section 3.8) and the project's risk profile (see Section 11.4). An understanding of these will assist the team in recommending the most appropriate procurement route for the building.

Of crucial importance: any construction project needs to align with the business plan, not simply at project level but also at programme and strategic levels. For this reason it is necessary to include in the value and risk management policies guidelines on upward referral of risks and value-adding opportunities through the value chain.

# **6** Concepts, standards and qualifications

## Summary

This chapter explores the underlying concepts of value and risk, the standards that exist and qualifications that are available, to give the reader a fuller understanding of the subjects.

- *Value*

  o First it explains how value is subjective and describes three fundamental categories of value: utility, exchange and esteem value.

  o It outlines the concepts of hard and soft values.

  o It introduces and explains the concept of value propositions, to describe an organisation's value priorities towards its customers.

  o It describes how the balanced score card (BSC) gives a method of assessing business performance by using metrics that look into the future as well as past performance. It explains how this can be adapted for use on projects.

  o It explains the concepts of value drivers and gives examples of possible generic value drivers for comparing similar building types.

  o The chapter goes on to explain how the National Audit Office (NAO) are using these concepts to assess the value of a project from a broader perspective than cost and time, based upon value drivers and good design.

  o It describes how value drivers can be used to articulate an unambiguous and clear project brief.

  o It introduces the output-based design quality indicator to assess the quality of design.

o    It describes how value can be represented as a balance between benefits and investment, and the importance of whole-life costs. It introduces the concept of whole-life value.

o    It summarises the British Standard for value management, BS EN 12973:2000 and the main principles expressed therein.

■    *Risk*

o    The chapter describes the concepts of risk and how risk may be viewed as the flipside of value.

o    It goes on to introduce different types of risk and how the occurrence of project risks can trigger impacts outside the project that may affect the business.

o    It introduces commonly used categories of risk as an aid to their articulation, assessment and management.

o    The chapter goes on to explain the need to take risks to maximise value and the concepts of risk appetite and willingness (or aversion) to taking risks.

o    It outlines the principles of taking and controlling risk to maximise gains and explores the risk profiles associated with different procurement routes.

o    It summarises the principal messages contained within the risk management section of the Project Management Standard, BS 6079.

■    *Training and qualifications*

o    The chapter concludes with a resumé of the systems available in the United Kingdom to gain qualifications in value and risk management.

## 6.1 Understanding value and risk

The fundamental reason for undertaking any project is to add value in the form of benefits to the organisation(s) commissioning the project. To do this it is necessary to articulate what value means to that organisation in terms that are unambiguous and easily understood by all those who are involved in realising the project. Ways of doing this are discussed in Chapters 2 and 10. Doing it well requires an understanding of value.

There is no such thing as a risk-free project. To manage the risk requires the processes that are described in Chapters 3 and 11. It is not enough to simply apply processes to control risks effectively. It requires a deeper understanding of risk. This chapter provides a basis from which to understand value and risk.

## 6.2 Understanding value

### *Concepts of value*

Value is one of those things that everyone understands but which no two people will describe in the same way. You cannot touch it or (easily) measure it. And yet it is one of the most powerful concepts in the market place. Frequently, suppliers are told that their quotations to provide a product or service will be judged according to value for money. Ask the person sending out that enquiry what they mean and you seldom get a clear answer. Ask two people and you will get two, different, answers.

### Differing stakeholders perspectives

To manage value, the team must do better than this. They need to be able to define value in terms that everyone can understand and measure, at least qualitatively. This chapter introduces some of the underlying concepts of value. Later it describes how to use these concepts in value management.

Different people view value differently. They have different values. Their views will be coloured by their personal taste, the impact of the thing they are valuing, their circumstances and a host of other factors.

Value is subjective. This also provides the first challenge in value management: how to measure something which is intangible and subjective. If we cannot measure it, how can we manage it?

Early pioneers of value management identified three factors influencing value:

■   Utility – will it work effectively and do what it is designed to do?

■   Exchange – can I sell it for a profit?

■   Esteem – will it convey status or make me feel good?

The relative importance of these three core types of value will vary depending on an individual's values.

## Utility value

A factory unit is built for one overriding reason. It must enable the efficient manufacture of a product. If it does not do this well it will be deemed a failure. An engine must deliver power reliably. If it does not fulfil this function effectively it will be of little use.

In these examples, the utility of the system is of primary importance. There may be other, secondary components of value. For example, not only an engine must deliver power but also do it reliably.

Most buildings are constructed in order to accommodate and support specific activities. The building will be judged a failure if it does not do this effectively. Thus, maximising the productivity of what is done within it is a key component of the utility value in many buildings.

In Chapter 2 we saw how optimising the design of a facility increased productivity for a bank by 23%, thus paying for the development within less than a year.

## Exchange value

The property market is driven by the concept of exchange value. Developers have one primary goal – to maximise the value of their asset, either through a sale to a third party or in terms of future rental income. Both routes rely on exchange – ownership or tenancy of the building in exchange for money.

Exchange value relies on the fact that the parties involved in the exchange have different values. The concept of value drivers enables project teams to optimise value for their projects. Usually this will involve trade-offs (exchanges) between different stakeholders to obtain the optimum balance between their differing values.

A well-designed building, offering a productive and pleasant environment to its users, will be far more saleable than one which is built for the lowest cost and possesses neither of these qualities. The architectural excellence for which a Chicago AIA Award was given in 1981 was reckoned to add $1.64–1.85 per square foot to the annual rent (or about 10% in those days) (refer Hough and Kratz 1981/2).

Another example of realising exchange value was provided by a lift manufacturer who, instead of offering standard lifts, offered a turnkey package to provide 'vertical transportation' for the contract period in a private finance initiative (PFI) project. Under the deal, the lift maker operates and maintains lifts and incurs non-availability penalties (i.e. shares risk and reward). The building users pay for the outcome they seek, vertical transportation.

## Esteem value

Esteem is a primary value for buildings that need to convey an image or otherwise contribute to their environment. Corporate headquarters must convey to the public and clients alike what the corporation is about – that it is successful, it cares about details and it cares about its customers. In short, the building must convey the corporate brand. Of course, it must work as a building (utility value) and it must be saleable as an exit strategy (exchange value), but of overriding importance is the esteem in which the outside world hold the corporation.

The plan form of National Westminster Tower in the City of London (now Tower 42) was built in the shape of the bank's logo to convey the power of the brand to the public.

The Eden Project brings over £150m a year in contingent inward investment to Cornwall.

The Deep Aquarium, Hull's £45.5m Millennium Commission lottery project is not only a wonderful piece of architecture, designed by Sir Terry Farrell, but is a tremendous economic boost to the area. It functions as a tourist attraction, a lifelong learning centre, a business centre of excellence and an important research centre run in conjunction with University of Hull. Signifying the changing face of the city, it is the first physical realisation of the massive urban regeneration taking place in Hull.

## Hard and soft values

When considering alternative engineering solutions for building components there is usually a single preferred solution. It either does what is required or it does not, even though it may present opportunities for improvement. Such value assessments may be labelled 'hard'.

Issues involving people can be more complicated. In resolving the optimum working environment between a number of stakeholders with differing needs, there may be no single right answer. What is perfect for one group may be unacceptable to another. It is

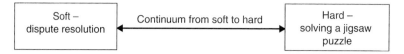

**Figure 6.1** The value continuum

necessary to discuss the situation with all the stakeholders and come up with a consensus solution that all can accept. In these circumstances, the value assessments may be labelled 'soft'. There is no single 'right' solution.

In his paper to the Hong Kong International Conference in 1998 Professor Roy Barton proposed the notion of a continuum between hard and soft problems. At one extreme is a hard problem like solving a jigsaw puzzle (Figure 6.1). There is a single right way to do this and, once done, the puzzle is either complete or it is not. At the other extreme is the ultimate soft problem, finding a solution that is acceptable to all parties in a long running dispute. Such a problem may remain unresolved for long periods of time, despite the best efforts of many highly intelligent people. Even when a 'solution' is found, there will be some who find it hard to accept. Between these two extremes are various degrees of hard and soft. The value judgements needed to resolve these problems span the same continuum.

Soft value management skills are used more in the early project stages when the project is not fully defined. This usually involves reaching consensus with many different stakeholders. As the design develops towards resolved design solutions, so hard value management skills and methods increase in importance.

## Value propositions

While the above classifications of value may apply to individuals' perceptions of value, different organisations will also adopt different value propositions in order to allow them to compete effectively in their business environment. The concept of value propositions was evolved by Treacy and Wiersema in 1993. It was born out of a desire to understand why certain businesses thrived and why some failed, even though they were operating in similar business environments.

The concept takes us away from traditional market segmentation, based on product type, geographic region or socioeconomic group. Rather, it revolves around three specific value propositions based on what makes your company the customers' first choice.

**Figure 6.2** Value propositions

Recognising this and adopting an operational structure to deliver it is the key to success.

All customers are different and attracted by different product characteristics. They have different values. Treacy and Wiersema distilled these different characteristics down to just three different value propositions (Figure 6.2):

■ Product leadership

■ Operational excellence

■ Customer intimacy.

Delivering each requires a different set of business processes and, consequently, different skills, management and information systems. They also require different values and beliefs and, thus, a different value culture in the organisation. These are illustrated in Table 6.1.

To be successful, an organisation should be competent in all three value propositions but it should excel in just one. To attempt to excel in all three requires three conflicting operational models. This is clearly not practical and is likely to lead to mediocre performance in all three. Further reading is indicated in the references in Appendix A.

For a building to be a success, it is necessary that all members of the project team understand the chosen value proposition of the organisation with whom they are working. The building should reflect this. The value study leader may include such an analysis to help inform the decision-making process.

**Table 6.1** Value proposition characteristics

| Value proposition | Characteristic | Customer attraction | Example | Business processes |
|---|---|---|---|---|
| Product leadership | Innovative; setting the trend; leading edge; must have, stylish | Innovation and trendiness more important than cost, reliability or after-sales service | Software corporations, telecoms | Research and product development; innovation; understanding the market |
| Operational excellence | A good deal, low cost, no frills, reliable | Good value for money; price and reliability are the main drivers | Low cost, no frills airlines | End to end supply chain optimisation; emphasis on reliability and efficiency |
| Customer intimacy | Bespoke service; attention to personal needs | Personal service inspiring great loyalty; not the cheapest | Any good restaurant; bespoke tailors | Customer service; understanding the market; flexible and responsive |

## *Articulating and assessing value*

### The balanced score card

If value propositions provide a picture of an organisation's chosen stance towards its customers, the BSC provides management with a forward looking set of data to monitor their performance.

Robert Kaplan and David Norton developed the concept of the BSC in the early 1990s. They did so in response to the notion that, while traditional financial measures were adequate for organisations that evolved relatively slowly, they provided measures only of what had happened in the past. They provided no guide as to future performance, which is seen as necessary for contemporary, fast-changing organisations.

The BSC provides information on four basic fronts (Table 6.2):

■ How do our customers see us?

■ What must we excel at?

■ Can we continue to improve and create value?

■ How do we look to our shareholders?

**Table 6.2** The principles of the Balanced Score Card

| Perspective | Subject addressed |
|---|---|
| Financial | How do we look at our shareholders? |
| Customer | How do our customers see us? |
| Internal business | Can we continue to improve and create value? |
| Innovations and learning | At what must we excel and how can we do this? |

## Wide consultation

Experience with the BSC has shown another key advantage, wholly in line with the value-based culture encouraged by value management. In order to gather and interpret the operational information (reflected in the measures for customer, internal business, and innovation and learning perspectives) it is necessary to involve all the management and the staff, not just the financial controllers. This wider involvement helps all members of the organisation to understand their goals better and build greater team collaboration. In effect, it puts the management strategies at the heart of the organisation and encourages the necessary employee behaviours to implement those strategies.

## Project application

It is possible to redefine the value perspectives of the corporate BSC shown above to reflect the priorities in managing building projects. Table 6.3 proposes equivalent headings, along with possible goals and measures.

In Table 6.3 we have introduced some examples of what goals and measures might be used to build a BSC. The completed picture provides all relevant information in a single report, enabling management to address areas of potential weakness and thus to optimise their performance.

## Value drivers

In fact, there are generally a number of components of value, each of which contributes to the overall value of the item in question. These may be described as the value drivers – those things that, taken together, include all the benefits that contribute to the value of the completed project to the business. If all of the value drivers are delivered successfully there should be nothing else to do to deliver a successful project. But if any of the value drivers is not delivered in full, the project will fall short of its objectives. Achieving the value drivers therefore is vital for success.

**Table 6.3** Building project 'Balanced Score Card'

| Corporate BSC perspective | Building project BSC perspective | Possible goals | Possible measures |
|---|---|---|---|
| Financial | Business objective | Achieve 20% profit | Financial reports |
| Customer | Stakeholder | Public acceptance | Planning consent |
| Internal business | Programme and project management | Complete on time and budget | Monthly progress reports |
| Innovation and learning | Team competencies and skill development | Qualified staff | Training and certification |

## Outcome focus

Value and risk management are outcome focused. There is therefore a requirement to develop measures of performance based on outcomes. Value drivers provide a way of doing this.

The generic family of value drivers illustrated in Table 6.4 apply to most building types. The first two of these relate to process, the rest to outcomes.

## Benchmarking by value

Using Table 6.4, it is possible to derive common value drivers for different types of building.

An organisation that specialises in one or more specific building types can customise these value drivers accordingly. It can then benchmark buildings by value, as distinct from more conventional benchmarks related to cost and time for delivery. Table 6.5 illustrates value drivers customised for educational buildings. The example shown in the table is an extract from a generic value profile for educational projects.

## Functions and benefits

Readers who are familiar with the language of value management would have recognised the above descriptions of value drivers as functions.

Senior management tend to understand the term 'value driver' more readily than the term 'function'. In commonly accepted value management terminology, value drivers are described as primary functions. Functions explain what things must do, what they contribute to the whole. The importance of considering the function of a component part of a system is central to the management of

**Table 6.4** Generic value drivers

| | Value driver | Addresses the question |
|---|---|---|
| 1 | Enhance/achieve desired financial performance (of the building) | Is the building affordable? |
| 2 | Manage the procurement process effectively (maximise project delivery efficiency, minimise waste) | Are the project management processes efficient? Are the right people engaged? Is the supply chain effectively managed? Are materials and labour used efficiently? |
| 3 | Maximise operational efficiency, minimise operational costs | Does the building work well for the end users? |
| 4 | Attract and retain employees/occupants | Is it a nice place to work? |
| 5 | Project the appropriate image | Does the building convey the appropriate image? |
| 6 | Minimise maintenance costs | Is the building easy to maintain and clean? |
| 7 | Enhance the environment | Is the building environmentally friendly? Is it built using the principles of environmental sustainability? |
| 8 | Comply with third-party constraints | Does the building conform to legal and other external stakeholder requirements? |
| 9 | Ensure health and safety during project implementation and in operation | Is the building safe to build and use? |

value. It is explained more fully in Chapter 9. Functions explain what things *do*, in terms of contributing to the utility, esteem or exchange values, rather than what things are. This paradigm shift in thinking (for many people) is what adds value.

## Good design

The Government's Better Public Building Initiative in 2001 showed concerns that auditors of the National Audit Office (NAO) tended to base their assessments of public buildings with too much emphasis on costs and time and not enough on the qualities of good design. This attracted the interest of the Commission for Architecture in the Built Environment (CABE) who recommend the use of value drivers to assess the quality of well-designed buildings and their ability to deliver the benefits for which they were intended.

**Table 6.5** Extract from an educational project value profile

| | Generic value drivers | Examples of value drivers for educational building projects | Comments |
| --- | --- | --- | --- |
| 1 | Business and financial | Comply with funding requirements<br>Obtain funding and approvals<br>Achieve commercial viability<br>Generate revenue<br>Enable research funding to be obtained<br>Increase teaching capacity<br>Involve local community and third parties | These value drivers relate to the need to obtain funding whether this be from public sources or commercial activities. The built development needs to satisfy funders' requirements and allow for revenue generation |
| 2 | Delight staff, students and other stakeholders | Inspire learning<br>Create delightful space<br>Make it a positive experience<br>Be recognisable as international icon of learning<br>Attract best international students | Education buildings are important public buildings used by many people they should delight and inspire |
| 3 | Maximise operational efficiency | Improve interaction between departments<br>Optimise use of space<br>Deliver functionality to match curriculum<br>Enable flexibility to change future use<br>Control circulation<br>Optimise adjacencies<br>Provide privacy where needed<br>Facilitate multidisciplinary activity<br>Create a rich and stimulating environment<br>Provide a good working environment for users<br>Create 24 h access<br>Integrate new building within existing estate<br>Raise levels of aspiration in students<br>Encourage innovative management and teaching | The building needs to be practical to operate efficiently. It is also of key importance that the building actively enables the work and activities of the organisation, perhaps, by offering opportunities for communication and interaction |

## Articulating value

The Re-thinking Construction initiative has had a dramatic effect on improving performance in our construction industry. Demonstration projects using the principles of Re-thinking Construction claim that projects that follow the principles can result in capital cost reductions of 10% or more and result in other benefits such as reduced defects. In the main, these improvements arise from gains in the efficiency of the construction process. However, this is little to ensure that the right building is created to deliver the full benefits expected by the client.

## Clarity of brief

It is widely recognised that developing a clear brief and conveying this to those who will implement the project is one of the most difficult tasks faced by the team. More projects fall short of their ideals due to misunderstandings or changes in the brief than for any other single cause. One of the reasons for this is that there is no simple and clear method for articulating a clear, unambiguous brief that fully reflects the client's requirements.

Some briefs are so obscure that it would seem that the writers wish to emulate Alan Greenspan, the veteran American chairman of the Federal Reserve, who once said 'if you think you have understood what I said then I have obviously not made myself clear'.

## Lowest cost

Suppliers are often told that their bids will be assessed on a basis of value for money. In fact, the assessment criteria usually comprise a number of statements to assess whether the supplier is capable of delivering the service or materials to the quality and specification laid down in the invitation documents.

Since most competent suppliers can conform to these requirements, the choice comes down to lowest cost, not value for money.

So, what is needed? Most importantly, a reliable way for the commissioning authorities to articulate the long-term value required from a building, for the supply side to be able to respond with propositions that demonstrate how they will add value and for the buyers to be able to take full account of such value in their decision-making process.

## The value cascade to inform the brief

Value management provides one very effective means of doing just this. Every building project should have clearly stated objectives, expressed in terms of the expected benefits. Value management breaks these objectives down into a number of functions (or value drivers). Here, the term function describes what things must *do* in order to contribute to the objectives. This is a fundamental shift in thinking from most specifications in which things are described by what they *are*.

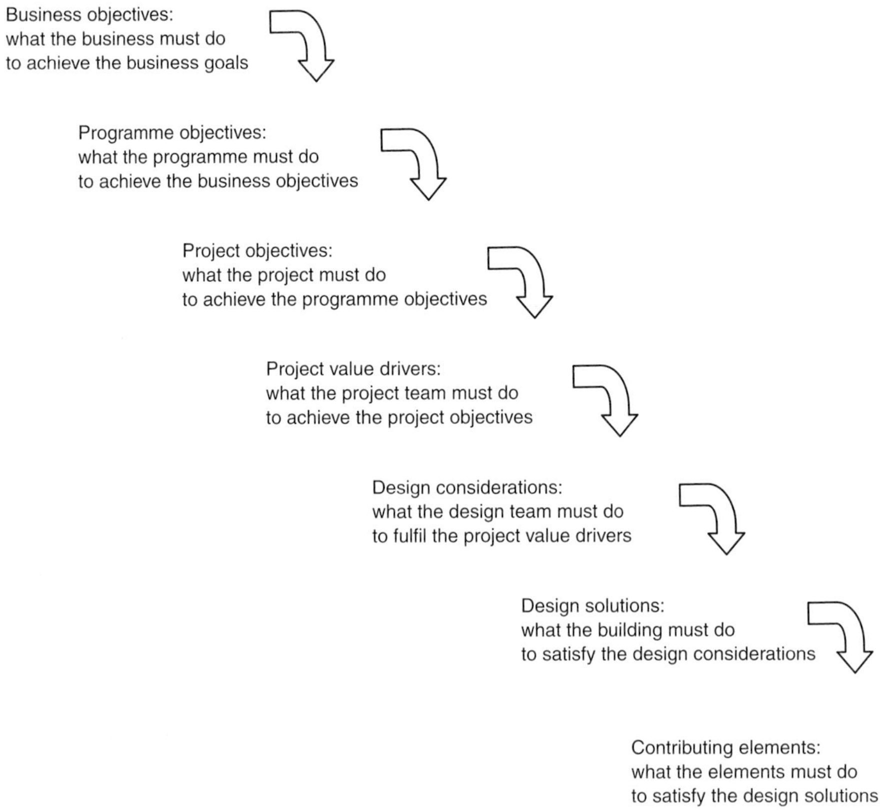

Business objectives:
what the business must do
to achieve the business goals

Programme objectives:
what the programme must do
to achieve the business objectives

Project objectives:
what the project must do
to achieve the programme objectives

Project value drivers:
what the project team must do
to achieve the project objectives

Design considerations:
what the design team must do
to fulfil the project value drivers

Design solutions:
what the building must do
to satisfy the design considerations

Contributing elements:
what the elements must do
to satisfy the design solutions

**Figure 6.3** The value cascade

The project must deliver on all value drivers, in full, if it is to be a complete success. The relative importance of each value driver will depend upon the client's priorities (e.g. in some buildings, user productivity may be the all important value driver, while for others, projecting the appropriate image may predominate).

The project team build a model, linking the objectives, the primary functions or value drivers, the resulting design considerations, the proposed design solutions and selected elements of construction. Figure 6.3 illustrates the principles of the resulting model which we call the value cascade. The team can descend the cascade by asking the question 'how?'. They can ascend it by asking the question 'why?'.

Developing a comprehensive model, similar to that illustrated above, forces all members of the team, including the client, to express their requirements in clear, simple language that all can understand. The resulting model provides a readily understood summary of the requirements of the brief, the expectations of the designers and the reasons why certain design solutions are preferred.

The same model provides a clear expression of what value means for the client and, therefore, a means for suppliers to articulate how they will add value in the execution of their services.

If, as frequently happens during the execution of a large project, the project objectives change, this would be reflected by changes within the value drivers throughout the cascade. This enables the client body to describe the changes in emphasis within the project that are needed to respond to the changes in project objectives.

## Design quality indicator

Further support for the emphasis on good design is provided by the Construction Industry Council in the form of the design quality indicator (DQI). This is a method of assessing the design quality of buildings. It is a somewhat narrower measure than that used by the NAO, since it places less emphasis on the fitness of the building for the purposes of its users. The DQI focuses on the outcome of the project of the completed buildings rather than the process of delivering it and, therefore, is consistent with value and risk management thinking. The method is designed to be used at any stage during the development of a building design and relies upon the project stakeholders completing a 90-point questionnaire. Questions address the three fundamental quality fields of build quality, functionality and impact.

At project inception, use of the DQI enables the client's end users and other stakeholders to define their requirements in terms of the quality of the finished building. It contributes to the understanding of client and user requirements and their preferences and assists in the development of the project brief and the understanding of trade-off between different choices. On PFI schemes it is a useful way to amplify the outputs specification and assess bidder responses.

At design stage, the DQI can help focus on areas which could be improved or where requirements are exceeded and could possibly be relaxed.

During construction, the tool allows all those involved in procurement to convey the project's aspirations to the supply chain.

Once the building is complete and in use, the DQI can be used to evaluate users and the public's perceptions of the building

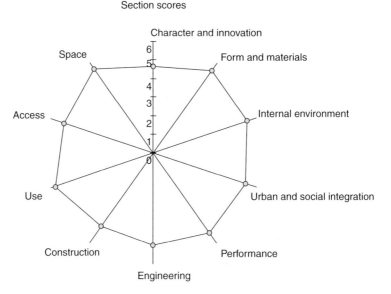

**Figure 6.4** DQI for British Library Centre for Conservation

at post-occupancy stages. Outputs from the DQI are normally displayed by means of spider diagrams as indicated in Figure 6.4.

Further information may be obtained at the DQI website.

## Optimising the balance

Achieving the value drivers in full may ensure that the project delivers all the required benefits but at what cost? To spend too much money to deliver the benefits risks undermining them.

## Perfection comes at a price

Perfection comes at a price but spending too little risks falling short of objectives. Clearly there is an optimum balance. Just the right amount to enable completion of the project objectives in full with nothing left undone and no waste. It is not just cost that may overrun. Achieving perfection may require more time than is desirable. It may require more of a scarce commodity than is available. In other words, delivering the required benefits may consume more resources (cost, time and materials) than is desirable. This goes to the heart of delivering best value, finding the optimum balance between the use of resources and the benefits gained. Chapters 8 and 10 expose the reader to the tools and techniques that enable a team of people to arrive at the optimum balance. Note that in the context of value management and the government's best value policy, the term value refers to the optimum balance between perfection and affordability.

## The value ratio

The value management standard, BS EN 12973:2000 describes value as being proportional to the ratio of the 'satisfaction of needs' and the 'use of resources' or

$$\text{Value} \; \alpha \; \frac{\text{Satisfaction of needs}}{\text{Use of resources}}$$

(the symbol $\alpha$ means 'is proportional to').

People's values are what determine the upper line in this relationship. These are dimensionless and have nothing to do with money or other resources that may be used in order to realise them. These values will vary between people. They will also vary with the circumstances in which that person finds himself. The bottom line comprises all the resources (e.g. time, cost, materials and manpower) needed to satisfy the needs. The above equation is known as the *value ratio*.

The value ratio is frequently presented by the narrower expression:

$$\text{Value} \; \alpha \; \frac{\text{Function}}{\text{Cost}}$$

The degree to which needs (or project objectives) are satisfied may be described by reference to the value drivers or functions. The use of resources often comes down to cost, but may concern other resources such as time, which cannot be directly related to cost. BS PD 663:2000, the Guidelines to The BS EN 12973:2000 simply express the value ratio as

$$\text{Value} \; \alpha \; \frac{\text{Benefits}}{\text{Investment}}$$

## Whole-life costs

The cost of undertaking the project is only a part of the total cost of a facility to the owner-occupier. Once it is in use, there are the costs of operating the building and maintaining it in good condition. On top of these are the costs of conducting the enterprise for which the building was built. The latter can outweigh the initial construction costs many times over.

In any exercise to maximise value, account must be taken of the whole-life costs. If capital costs are cut and, as a result, running costs increase, even by a small amount, the effect is likely to be a reduction in the net present value of the project. On the other hand, more expensive equipment may be cheaper to maintain and run.

A well-known car rental company has standardised on a fleet of A-Class Mercedes cars. They are certainly not the cheapest cars in the market to buy but they are economical to run and maintain. In the long term the savings in running costs outweigh the higher initial purchase costs, thus making the cars cheaper overall. This enables the car hire company to charge less for hiring its cars than the competition. It gives the company competitive advantage. By spending more capital the car hire company has gained competitive advantage, or added value, through reduced running costs.

The same principle applies to building projects.

## Whole-life value

The UK government's move from cash-based accounting to resource-based accounting in 2001 gave the central and local government the incentive to invest, since it takes account of whole-life costs and benefits. Whole-life value (WLV) is described as the 'assessment of the benefits and costs of an investment over time, taking account of stakeholder interests'. It seeks to reflect the, so-called 'triple bottom line' reflecting economic, environmental and social balances of sustainability.

Barriers to achieving optimum WLV include

■  Inertia (resistance to change)

■  Leadership (including lack of common voice from institutions)

■  Lack of robust mechanism to transfer costs and benefits

■  Definition and methodology (value and risk management can help here)

■  Short-termism

■  Inappropriate procurement mechanisms (e.g. wrongly focused incentives, for example, where users pay for energy do not encourage efficient design)

■  Lack of innovation (another area where value management can help).

Risks to being able to deliver optimal WLV include:

■  Change

■  Obsolescence

- Build quality

- Maintainability quality

- Design.

## *Value management standards*

In the early 1990s, the European Commission recognised that value management had an important part to play in improving national productivity. It therefore promoted a number of programmes to improve understanding of value management across Europe. One of these was that CEN (European Standard Organisation) commissioned the writing of a value management standard.

At about the same time, the Institute of Value Management introduced a strategy to uphold the quality of value management service delivery since it was perceived that poor quality delivery was undermining confidence in the discipline. One leg of this strategy was to develop a standard in value management, the second was to use the standard as the basis for a training programme and the third was to introduce certification of experienced practitioners (Figure 6.5).

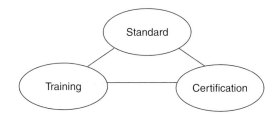

**Figure 6.5** Strategy to uphold the quality of value management practice

### The British and European Standard in value management

The principles of value management are described in the British Standard BS EN 12973 : 2000 (developed in 1994–5 but not published until 2000). The document is imperfect in that it represents views of 14 value management associations across Europe and includes a number of compromises. It does, however, set out a number of powerful principles. Because the document is not an easy read the British Standards Institution (BSI) published PD 6663 : 2000, a guide to the BS EN 12973 : 2000 in the same year. This is in effect an interpretation of the full standard into much more easily understandable English and a précis of its contents. These two documents now underpin best practice in modern value management in the United Kingdom.

Because it had its roots in improving productivity, the standard addresses value management from two standpoints. First, from the point of view of introducing the principles of value management into a business and, second, the application of value management to projects. Any organisation wishing to use value management regularly needs to evolve a value culture which aligns with the aims of the business if maximum benefit is to be obtained. This process is described in Section 4.9.

## The IVM/European training and certification system

Based upon the standard, the European Commission sponsored the development of a comprehensive Training and Certification System in value management. This was launched in 1997. The system is based upon nine primary learning objectives, all of which are necessary to deliver the primary objective of the system, namely to impart knowledge, skill and attitude (i.e. competence) to individuals in the delivery of value management. The resulting qualifications provides the means for successful individuals to demonstrate competence.

The subjects covered by the primary learning objectives are:

1.  Human dynamics

2.  Communications with internal and external customers

3.  How to set up a project

4.  Expertise in the use of value management methods

5.  Manage interfaces with other (non-value management) projects and activities

6.  How to interact with business objectives

7.  Continuous improvement based on experience

8.  Execution of value management projects

9.  Development of a value culture.

The training courses comprise three levels:

1.  Awareness courses for all levels of management and employees to build a value culture.

2.  A three day foundation course to provide essential knowledge for aspiring practitioners.

3.  A six day advanced course to enable the candidate to lead value management studies and acquire the qualification, Professional in Value Management (PVM). This course may be divided into two parts, each of three days duration. After the first part, candidates are eligible to apply for the qualification Certificated Value Analyst (CVA), indicating that they are competent to run simple value analysis or engineering studies.

The qualifications require the candidates to demonstrate experience in actually running successful studies. Qualifications are assessed and awarded by the independent Certification Board of the Institute of Value Management. Further information is available from www.ivm.org.uk. The qualifications assist customers to assess the competence of individuals in delivering value management services.

## Value management principles

The value management standard names a number of principles of value management (Figure 6.6):

*Focus on outcomes.* The first of these, a focus on outcomes, has been mentioned already. Value management focuses more on the outcome of projects than the means by which those outcomes are delivered.

*Management style.* The second is a distinct management style focusing on the needs of the customers and an emphasis on the performance levels that they expect. It involves creativity and the pursuit of innovation, robust evaluation of ideas based on quantitative measures and an encouragement for people to work collaboratively together to solve problems. There is an emphasis on exploring the root causes of problems and resolving them, rather than treating symptoms.

*Positive human dynamics.* The next principle is that of positive human dynamics in which people are encouraged to work together in a proactive manner. There is a recognition that all individuals can make effective contributions, that the status quo is there to be challenged to bring about improvements, improving shared understanding of the problems which are faced. The methods encourage group decision-making leading to consensus and ownership of proposals by those who are responsible for implementing them.

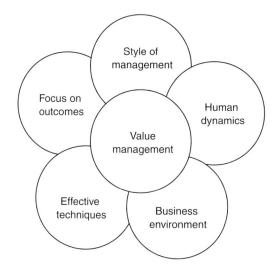

**Figure 6.6** The principles of value management

*Recognition of business environment.* The fourth value management principle involves recognition of the environment in which the organisation is conducting a project. There are certain pre-existing conditions over which the project team may have little or no influence (these are frequently known as 'givens'). The team must also recognise that what they do will have an influence on things outside their immediate project. At the same time, external influences may have an impact on their project. These are frequently referred to as the opportunities or constraints identified during a value management programme.

*Effective techniques.* The fifth principle is the presence of an effective toolkit of robust and proven techniques with which to achieve results.

## The American Standard

The Society of American Value Engineers (SAVE) published the equivalent of the American Standard in value management under the title of 'Value Methodology Standard' in 1997. The principle difference between this document and BS EN 12973 is in emphasis. The Value Methodology Standard contains less emphasis on human dynamics and management style and more upon the core technique of value engineering. The Value Methodology Standard is available through SAVE at www.value-eng.org.

## Cultural differences

Approaches to value management need to take account of differing cultures. These can arise between different organisations, different operating sectors or different countries. There is a well-known bank that promoted itself as understanding different peoples' needs by a series of advertisements highlighting the differences between accepted behaviours in countries around the world. Such understanding is important when operating in differing cultures whatever business is conducted.

> Early value engineering studies in the 1980s on defence instal-lations for the Property Services Agency (PSA), on behalf of the US Department of Defense, required the use of an independent design team to critique the proposed designs. This was the norm in the United States, where it appears design teams were used to having their designs scrutinised by third parties and would accept the independent recommendations. In the United Kingdom, the approach was highly unpopular. The incumbent design teams resented being told how they should have designed something by an independent team who had been involved in the design only for the duration of the value engineering study (about five days). The outcomes were counterproductive since the incum-bent team would, as a matter of principle and professional pride, devote considerable energies into demonstrating why the propos-als put forward by the value engineering team were inferior to their original designs. Apart from the bad image this gave value engineering, it also meant that good proposals were thrown out along with those of dubious merit. It did not take long before this confrontational approach, totally out of keeping with British pro-fessional culture, was replaced by the current practice of working with the incumbent design team.

In South East Asia there is a great reluctance for professionals to be seen to criticise their peers. The value management process requires the team to generate alternatives to existing proposals. There can, therefore, be a great reluctance to generate alternatives for fear that they may be seen as criticism. In these circumstances, it is necessary to explain carefully that the intent of generating alter-natives is to assist the original designers to improve on their original ideas, not to criticise them.

Similar cultural differences exist in different market sectors and between different disciplines. For example, architects tend to think very differently from most quantity surveyors who think differently from engineers. Not appreciating such cultural differences can adversely affect the smooth running of a study. It is up to the study leader to make a point of understanding the culture of the people with whom he will be working so that he can anticipate and overcome any problems before they arise. He should do this as part of the preparations before a study.

## 6.3 Understanding risk

### *Concepts of risk*

Risk is present in everything that we do. We can never quite be certain that things will turn out exactly as we expect or hope. In everyday language, risk can be defined as the possibility that something unexpected and, usually, unpleasant will happen. Whatever we are doing could be dangerous or turn out worse than we had expected. But risk can also have a positive side. If we view risk as uncertainty of outcome, then things could turn out better than we expect.

The fact that there are opportunities associated with risk provides a direct link with value. Risk is the flipside of value. If risk occurs and results in an outcome being worse than we expected, value will be lost. Conversely, if the outcome is better than we expect, the opportunity provides for an increase in value.

There is no such thing as a risk-free project. Some projects may be more or less risky but none will be free of risk. Risk is particularly rife in construction projects. One reason for this is that no two construction projects are ever identical. To a greater or lesser extent every construction project contains elements of the unknown, from innovative design, working with different people, unknown circumstances, and the fact that much of the building is made on site, where conditions can be far from ideal and therefore difficult to control.

Risks can take many forms. Many risks are likely to affect the cost, some will affect time, others will affect quality. Some will threaten health and safety. Some will affect all of these and threaten the business. Different risks may be interrelated so that avoidance of one risk can trigger another. It is this variety and interrelationship of risks that can complicate what is, otherwise, a relatively straightforward process of risk management.

Risk is better understood as a concept than value. People continually assess risk as they go about their daily lives when they weigh up whether to take one course of action or another. It is, however, subjective. Some people are more risk averse than others. Some thrive on taking risks. Understanding the risk profile of individuals and the organisation to which they belong is, therefore, vital to those charged with managing it.

## Types of risk

There are many types of risks associated with a construction project, ranging from onsite risks during construction through to the business risks faced by the client. The focus of this book is on project risk management, however, those managing the project risks must be aware of the risks to which other stakeholders within the project delivery team including the client may be exposed. It is the duty of the managers of project risks to make the client and other stakeholders aware of potential risks arising from the project to which they may be exposed. The project may have in place the best risk management system in the world, but there is always a possibility that this fails to control the risk. When a risk occurs, it may have knock-on consequences far beyond the project itself.

> For example, consider a hospital project on which it is necessary to excavate a trench adjacent to the main power supply. Imagine that the hospital is in use and that power supply feeds the life support units. No doubt the contractor will have identified that severing the main power supply is a potential risk both to himself and others. As part of his risk mitigation strategy he would have taken great care to minimise the likelihood of the risk occurring. However, no safeguards are fool proof. It is possible that, because of something unforeseen, an inaccurate survey, perhaps, a digger does indeed sever the cable. Catastrophic as this may be for the digger driver, expensive as it may be for the contractor who will no doubt be held responsible and have to make repairs to the cable, imagine the consequences of the poor soul at the end of the life support system. Without any contingent plans made by the hospital authorities, he would surely die. This would be likely to lead to extreme adverse publicity. Public perception of both the hospital and the building contractor would suffer.

To avoid such incidents it is essential that the project team make the client aware of potential risks arising from the project. It is useful to think about these under three headings:

*Consequential risk.* This is the direct result of something happening on the project which disrupts the client's day to day operations. The consequences cannot be managed by the project team who have no

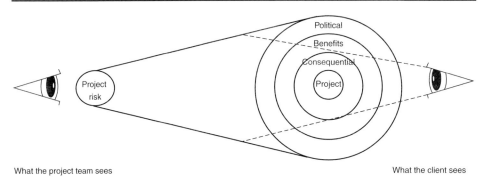

What the project team sees                                        What the client sees

**Figure 6.7** Types of risk

jurisdiction or authority outside the immediate project. The client body, however, provided they are aware of the possibility, can put in place appropriate contingency measures.

*Benefit risk.* During the design development of a project, it may not be possible to develop a facility that provides all of the expected benefits. For example, compliance with planning requirements may curtail the sizes of some elements of the facility. These could result in reduction in anticipated revenue. Once again the project team is powerless to eliminate this risk. The client, however, provided he is aware of it, can safeguard his business case by undertaking sensitivity analysis, modelling the potential loss in revenue and build-in allowances to enhance its robustness.

*Business and political risk.* Both consequential risk and benefit risk can be confined to the client body and be dealt with internally. However, there is a risk that the occurrence of one of these risks can break out into the public domain. For example, the loss of revenue may be so great that business is adversely affected. This could affect the organisation's share prices. Management needs to have plans in place to minimise such impacts should their first line of defence, project risk management, fail.

The project team tend to see only the project-related risks. In Figure 6.7, they view risks from the left-hand side of the diagram. The client, however, will be aware of the, much larger, consequential, benefits and political risks.

## Common reasons for project failure

Here it is worth dwelling on what we mean by project failure. Some projects fail so catastrophically that they are never completed. Here there is no need to define failure. What is much more common, however, is that the project is deemed a failure by certain stakeholders (not necessarily by all) because it has overrun on cost or time or failed to match up to expectations. Here we define project failure as a failure to deliver the full benefits expected of the project at the outset.

This includes cost and time overruns (because these will adversely affect the net present value of the project) and a failure to deliver the softer value drivers or benefits, such as image or user comfort. It will also cover any shortfalls in productivity in the facility which would adversely affect revenue streams.

The next section sets out a range of risk categories which relate directly to the main common causes of project failure. Provided these are managed effectively throughout the project, it is likely that the project will deliver success as defined by the value drivers.

## *Articulation and assessment of risk*

### Risk categories

Since the number of risks in any project is likely to be large, it is useful to break them down into categories. The Office for Government Commerce (OGC) Management of Risk (M_o_R) guidelines suggests the following categories:

- Strategic and commercial risks
- Economic, financial and market risks
- Legal, contractual and regulatory risks
- Organisation, management and human risks
- Political risks and those associated with society
- Environmental factors and acts of God (force majeure)
- Technical, operational and infrastructure risks.

Another approach is to use the common reasons for project failure, referred to in the previous section, as the basis for categorising risks. Table 6.6 lists these and may be used as the basis for developing generic risk management models.

There is no 'right' set of categories. The choice will depend on those managing the project and the business environment within which they are operating.

**Table 6.6** Common causes of project failure

| | Risk category |
|---|---|
| 1 | Funding and business issues |
| 2 | Clarity and understanding of client brief and objectives |
| 3 | Team roles, responsibilities and competencies |
| 4 | Management structure, lines of authority and communication |
| 5 | Schedule, information release, decision-making, timing and adequacy |
| 6 | Third party and external disruptions to operations |
| 7 | Planning, statutory approvals and health and safety |
| 8 | Construction, site conditions, ground and weather |
| 9 | Procurement uncertainties, cost, time and quality |
| 10 | Operational issues |
| 11 | Unresolved or unresolvable design issues |
| 12 | Contractor solvency, competency and general site management |
| 13 | Force majeure, natural or manmade disasters |

## Nothing ventured nothing gained

In Section 3.4 we noted the principle that taking risks is necessary in order to gain rewards and the close link between risk and value. In the entrepreneurial culture, those who take calculated risks generally reap the greatest rewards (provided they can control the risks).

Stephen Marks, chairman of the clothing chain French Connection stated, after a doubling of the share price 'if you don't get things wrong, you're not experimenting or trying'.

Government employees, normally considered the most cautious of people, are encouraged to take risks, provided they can be controlled. Michael Whitehouse, Assistant Auditor General of the NAO stated, in his introduction to the OGC management of risk guide 'Failure to take opportunities can be a huge risk in itself'.

Fund managers generally offer their clients the choice between risky investments (such as stocks and shares, where values can go up or down), or relatively risk-free investments (such as savings accounts in blue-chip building societies). In the former, the rewards have on average over the years exceeded even property in the return on investment and far outstrip the returns from the safer investments. There is a risk that the company fails and you lose everything. The building society investor however is most unlikely to lose his capital, but the rewards are likely to be much lower.

The property developer takes large and calculated risks in investing in some developments. Many property developers make a lot of money and are very successful. Others fail – high risk, high reward. The salaried employee in a large firm may never get rich but he is unlikely to become bankrupt – low risk, low reward.

## Risk appetite

Different attitudes to risk-taking can arise for a number of reasons. Past experience is a significant factor. Confidence and character is another. Ignorance of risk by one party may be seen as a willingness to take risks by another. Having clear procedures in place to control risk provides the confidence to take risks. Absence of a means of control has the opposite effect. When assessing risk it is necessary to understand people's appetite for risk. To know whether they are risk averse or risk-taking. Assessments of the same risk from people at opposite ends of this risk spectrum will vary widely.

Organisations should define their appetite for risk-taking so that all employees use the same measures. The most commonly used way to define risk appetite is to use the heat map or qualitative assessment matrix, described in Chapter 3. Risks above certain ratings must be treated in a defined way or, including a requirement to escalate them to a higher level of management. For example, by reference to Figure 3.4, the rule might be that risks ranking more than 13 500 should be referred to the management board.

The home office, on their construction projects, use a 5 × 5 risk matrix in which each square has a different rating. Their risk appetite is defined by reference to Figure 6.8.

## Change and risk

Willingness to accept risk is closely associated with the willingness to accept change. Those who are labelled as innovators tend to be much more willing to take on risk than those who are labelled non-adopters who are generally risk averse. This relationship is indicated in Table 6.7.

## Risk profiles

Different types of construction projects will have different risk profiles. The risk profile can be described by reference to the numbers of risks in the risk register of a given severity rating as indicated in Figure 6.9.

The risk profile for different stakeholders in a project will be affected by the procurement route selected. Each procurement route has different contract documentation. One of the key roles of the contract documentation is to allocate risk between the demand side (the client) and the supply side (the contractor). The types of

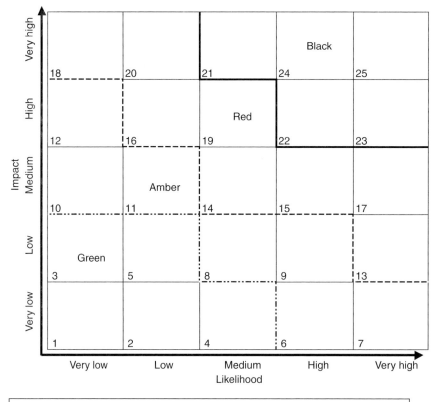

Figure 6.8 BRAG risk matrix to define risk appetite

risk that may be allocated through the contract are summarised in Figure 6.10.

The selection of the most appropriate procurement route for a particular project will be dependent upon the appetite for risk of the

**Table 6.7** Willingness to take risks

| Attitude to change | Behavioural characteristic | Risk appetite |
|---|---|---|
| Innovator | Takes up ideas willingly without external push | Risk-taking |
| Early adopter | Recognises the need but requires guidance | Risk neutral |
| Late adopter | Does not recognise need but will follow once convinced | Risk averse |
| Non-adopter | Resistant to change | Very risk averse |

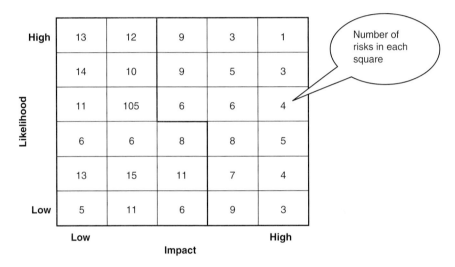

**Figure 6.9** Risk matrix showing risk profile

| Risks that no one can control (force majeure) | Risks that the client can best control (business related) | Risks that the client or contractor can control (allocated through contract) | Risks that the contractor can best control (project related) |
|---|---|---|---|
| Client owned | Client owned | Client or contractor owned | Contractor owned |

**Figure 6.10** Allocation of risk

organisation placing the contracts and should not be selected until the project team has established its risk profile.

## Risk management standards

There are several British Standards relating to risk management. Each relates to a different environment. Until recently, there was no standard that related to risk management in construction projects. At the end of the twentieth century, BSI published the project management standard BS 6079 of which the third part relates to the management of risks in projects. This document is not confined to construction projects and includes the management of risk in business-related projects. While an understanding of business risk is useful in successful project delivery it is not the subject of this book. The standard provides some very useful guidelines of best practice in the management of risk on construction projects.

It identifies several of the major causes of project failures. These include confusion over project objectives, the importance of stakeholder analysis and the profound impact than can arise from poor communications. These are consistent with categories shown in Figure 6.9.

## Risk management principles

The standard provides a generic framework for risk management which we reproduce in Figure 6.11 which is applicable in most situations including construction projects.

The standard defines certain terms used in risk management and these are reflected in the glossary of terms in Appendix B. It recognises that there are three levels of risk management:

1.    Strategic – relating to long-term issues

2.    Tactical – relating to medium-term issues

3.    Operational – relating to short-term issues.

This book is principally concerned with the tactical and operational levels of risk management although in some projects, for example, PFI, the project team will be concerned with the longer-term benefits and must therefore concern themselves with more strategic risks.

## Phases of risk management

The standard proposes that there are two broad phases of risk management. The first of these is concerned with defining the scope of risk to be managed and identifying the risks. The second phase is about assessing those risks and then managing them.

## Escalation

The standard requires that any risk management process should have a mechanism for escalating risks from one level to the next.

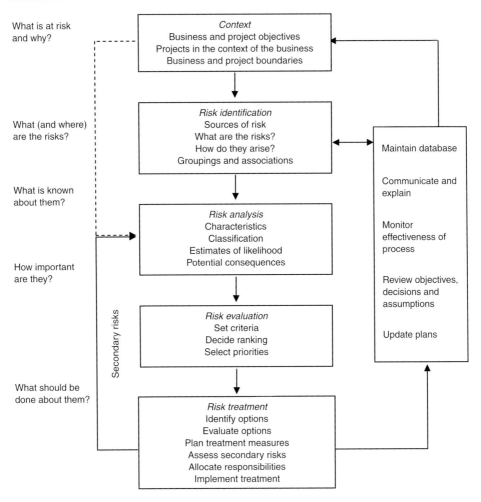

**Figure 6.11** A framework for implementing risk management

Thus a risk which begins at operational level and is directly related to, for example, a building project, should have a mechanism for identifying risks which it cannot control but which should be managed at tactical level. Likewise, risks being managed at a tactical level should, if the risk cannot be managed at that level, have a mechanism for escalating them to the next higher level the strategic level. This process is described in Chapters 3 and 11.

## Qualitative versus quantitative analysis

The standard emphasises the need for qualitative analysis for all risks and advises caution in proceeding to quantify risk. 'To what degree quantitative analysis is appropriate depends on the nature and quality of the data available for particular risks, the nature of the project, potential consequences, and whether analysis can provide additional useful information'. The standard is therefore quite clear that quantification for quantification sake is not recommended.

## Risk evaluation

The document proposes a simple qualitative method for evaluating risk by relating the potential effect of the consequence from the risk to the importance of the issue to which the risk relates.

## Risk management

Having evaluated the risks, the standard goes on to outline strategies for risk management and guidelines for their implementation. These guidelines are, unsurprisingly, completely consistent with the methods that we have described earlier in this book relating specifically to construction project risk management.

## 6.4 Training and certification systems

When a client commissions a building he is likely to engage a variety of professional advisors to help him maximise the opportunity and minimise the chances of things going wrong. These advisors will include legal, financial and agents as well as project managers, designers, cost consultants and others. Before engaging any of them, he will want to know that they are competent in their claimed fields of expertise. He will require that they demonstrate competence and track record among other selection criteria. We have seen that value and risk management can bring considerable benefits to projects, so it is sensible to check that those who are engaged to manage the process have the necessary individual competencies and track records to maximise the likelihood of success.

A good track record can only be gained over time and is easily checked by taking up references. Individual competence is supported by track record but should be underpinned by a recognised qualification. Many clients will state that they expect their professional teams to apply value and risk management (refer OGC Achieving Excellence Programme). Others will check that their advisors have the appropriate skills supported by qualifications.

One of the key reasons for writing this book is that there is not a lot of practical advice for clients or service providers on what represents best practice in value and risk management.

Frequently, the requirements for risk and value management are written into the scope of services required of the professional advisors, only for them to pay lip service to these duties and under-perform. Everyone loses through this practice. The owner, because he does not get value for money and may incur extra costs, time delay, or get a white elephant; the user, because the end product does not work well for him; the project delivery team, because they are involved in abortive work and fire fighting and therefore make less profit. Others in the supply chain suffer because they are denied the opportunity to make a positive contribution to the project.

There is an old saying that 'A job worth doing is worth doing well'. In manufacturing and the oil and gas industries, significant resource and effort is devoted to implementing effective value and risk management programmes, run by qualified and experienced practitioners. Often on construction projects there does not seem to be the same resolve and studies are frequently cut short and run by underqualified individuals. This partly due to the culture which exists in the industry, something that is being addressed through the Re-thinking Construction initiatives. It is partly due to the fact that appropriate qualifications are relatively recent developments.

There is good news. There are several internationally recognised qualifications for both value and risk management available in the United Kingdom.

## Qualifications

There are two primary sources of value management training and qualifications. The Institute of Value Management (IVM) in the United Kingdom has adopted the European Training and Certification System which leads to the qualifications CVA and PVM. The society of American Value Engineers have training and certification systems resulting in the qualifications Associated Value Specialist (AVS) and Certificated Value Specialist (CVS).

In 2005, the IVM and SAVE reached a reciprocity agreement under which people who qualified under the SAVE system can obtain IVM qualifications and vice versa.

There are two qualifications which are relevant to the management of risk of development and construction projects:

■ The Association for Project Management administer a training and certification scheme in which the practitioner level qualification follows some six days of training.

**Table 6.8** Value and risk management qualifications

| Accrediting organisation | System | Value and risk qualifications available | Acronym |
|---|---|---|---|
| Association for Project Management | APMP | Risk management: foundation and practitioner | APMP |
| APM Group Limited | OGC | Management of risk: foundation and practitioner | M_o_R |
| Institute of Value Management | IVM/European value management training and certification system | Foundation testimonial | None |
|  |  | Certificated Value Analyst | CVA |
|  |  | Professional in Value Management | PVM |
| Society of American Value Engineers | SAVE | Associated Value Specialist | AVS |
|  |  | Certified Value Specialist | CVS |

■ The OGC system, administered by the APM Group Ltd (not related to the Association for Project Management) issues a practitioner level certificate to successful candidates after about five days training.

This book is consistent with and contains most of the content of the syllabi for both of these courses.

This book is not the right place to set out the details of the above qualifications. Table 6.8, however, sets out the principal training and certification systems which are available and suitable for the construction sector and the awarding organisations.

*A word of warning!* Possession of a qualification on its own is not necessarily proof of individual competence.

## Experience

Some qualifications require minimum experience before the certificate is awarded. Others do not. In any event, the newly qualified practitioner, while he is able to demonstrate a level of competence, is not a 'seasoned professional'. This status can only be acquired by an individual who has been to practice successfully in a variety of project types at differing levels from strategic to operational, over a significant period of time. A newly qualified practitioner will be quite acceptable for many studies but some will require the additional experience of the seasoned professional.

Within larger organisation, it may be worth checking the maturity of their value and risk management processes.

## Qualified trainers

All of the courses identified in Table 6.8 require trainers to be accredited for experience and competence. Just because someone is good at what they do, it does not follow that they are competent to impart that knowledge and skills to a third party. Training requires specific competences to develop an appropriate training course, deliver it in a way that the participants not only gain knowledge but also acquire the skills and confidence to deliver the service. It is for this reason that trainers in value and risk management need special training (as trainers) if their trainees are to acquire the qualifications outlined above.

# 7 Learning from others

## Summary

- This chapter is devoted to guest papers, contributed by leading practitioners in their fields, outlining approaches to value and risk management and associated subjects to give the reader a broader perspective from the contributors' points of view. The papers included are:

    o A Value and Risk Approach to Project Development, contributed by Martyn Phillips

    o Soft Value Management, contributed by Stuart Green

    o Value Management in Manufacturing, contributed by Ken Scott

    o Value Management in Public Services, contributed by Clive Bone

    o Risk and Enterprise, contributed by Gerald Orman

    o The Impact of Partnering on Value and Risk, contributed by Mike Thomas

    o Maturity Modeling, contributed by Alan Harpham.

## 7.1 Practice in other sectors

### *Evolving best practice*

Previous sections of this book describe best practice in the application of both value and risk management to the construction projects. By definition, best practice must constantly evolve as people discover new and better ways of doing things and respond to an industry which is itself undergoing change. For most people best practice is something to which they aspire. It is not a description of common practice. The competent value and risk study leader will welcome such a change and contribute to it. He will evolve new ways of doing things without undermining the rigour of the processes upon which successful studies depend. To achieve the levels of competence necessary to do this, the study leader will need to accumulate experience in undertaking studies at all project stages over a wide variety of project types. He is likely to be a regular contributor to, and an active participant in, professional conferences and other gatherings to share his experiences and learn from others. In Chapter 5 we referred to such an individual a 'Seasoned Professional'. He will have great knowledge and skills but, nevertheless, never cease to inquire about new developments and learn new ways of enhancing his performance.

## *Guest papers*

One of the best sources of new learning is to observe how others do things, often in completely different sectors of the economy. To capture such experiences the author has invited a number of guest papers from the value and risk management community to contribute to this book, expressing their views on best practice in their own way. What follows is a précis of the contributions made by each of them.

## 7.2  A value and risk approach to project development

Contributed by Martyn R. Phillips, CVM, CVS, FICE, FCIWEM, P. Eng. Director, The TEAM FOCUS Group. Martyn Phillips is based in Canada and applies his value and risk management skills primarily to major infrastructure and energy projects, along with strategic performance alignment for business improvement. This paper, summarised from an article he wrote in the May 2002 *Proceedings of the Institution of Civil Engineering*, explains his approach.

The paper is based on experience particularly from within the fields of engineering and infrastructure. It would seem to be an implicit duty of project staff to endeavour to provide best value for money and satisfy stakeholders' other needs. But how does an individual know what will be acceptable to all stakeholders throughout the life of a project? Typically, most project team members focus inwardly. Some focus outwardly on the environment within which the project is being undertaken.

The integrated value and risk management team approach is a structured method for bringing together all different viewpoints to quickly identify choices for addressing complex situations and ensuring sustainable solutions at optimum cost. The following text provides an overview of applying the approach from early strategic assessment through concept development and implementation to optimisation of in-service facilities and routine operations. The focus is on improving project delivery and stakeholder satisfaction for increased cost-effectiveness and managed risk. A key aspect is consideration of life-cycle impacts.

## Introduction

The 'value methodology' has been practised for over fifty years. Key characteristics of the process are: early and continued stakeholder consultation; team alignment and culture change; a managed risk approach; consideration of whole-life impacts and integrated service delivery. Transparency of the process greatly aids decision-making and consensus development. With the formal inclusion of risk considerations, the value methodology is a particularly powerful project development aid and enabling mechanism for acceleration through the approval process.

## Context

Today, more than ever, there are many variables and viewpoints in any project, large or small. Value management provides a basic framework and a tool set that addresses issues of potential misunderstanding and misalignment at the start of a project. In many cases it results in significant reductions in the whole-life costs of the final project. The value and risk management approach is a natural companion to good programme and project management for complex or sensitive issue areas and for encouraging continuous improvement. It provides a vehicle for transforming the way organisations and individuals approach project planning and development. The methodology encompasses techniques to address the interrelated aspects of:

- Stakeholder issues and concerns

- Stakeholder values

- Project functionality

- Operations and maintenance requirements

- Costs (capital and whole-life)

- Implementation schedule

- Implementation obstacles

- Potential project risks

There are two contrasting applications:

1.  Strategic choice – the formulation of clear, unambiguous, strategic direction to enable approvals, funding and subsequent orientation of the development/implementation team. To build consensus on the way forward, through complete gathering of the many and various stakeholder views; strategic focusing is, of necessity, an iterative process.

2.  Value enhancement – through value engineering – continuing value improvement for finessing proposals to optimum quality, functionality and cost parameters.

## *Need for project improvement*

Construction projects in particular have acquired a reputation for confrontation between the contracting parties, and in some cases the public as well, resulting in:

- major claims and over expenditure

- delays and service disruption

- poor overall value for money

- stakeholder discontent.

The causes of these problems can often be traced back to misalignment of stakeholder expectations from the outset. To avoid such problems, there must be absolute clarity of context, needs, objectives and communications.

Value is determined not by the producer or promoter, but in concert with the customer and/or user. Nor is value related solely to money, as value criteria may include, for example, aesthetics, ease of operation and maintenance, environmental friendliness and provision for longer-term needs. Clients are really seeking to buy *performance*, not just traditional practices, or project development activities. Good project performance includes satisfying a wide range of stakeholders who may have differing views, values and thresholds of tolerance for perceived risk.

The value methodology is a powerful, group thinking process. The structured, formal study requires that a broad mix of stakeholder representatives focuses collectively on shared issues and opportunities. Outcomes are based on a common understanding of needs, constraints, key concerns, major risk areas, life-cycle impacts and shared/negotiated team values.

The value methodology objectively challenges assumptions, identifies alternative options, prioritises according to agreed criteria and then develops and tests the action plan for practicality of implementation. It is a very powerful, 'fast-tracking', consensus development tool. It has a tremendously synergistic effect, which overcomes otherwise adversarial relationships to develop a team approach and produce ownership of, and commitment to, the end-product.

To reach consensus, all participants need to have a similar understanding of each other's value and constraints. Workshop teams should be multi-disciplinary and represent the interests of all groups who may be impacted by the project under consideration. The mix of team members varies with the stage that the project is at, and whether an integrated team or an external, third-party team is used.

## Value study workshop process

Much of the power of the value and risk management methodology lies in the rigorous, disciplined approach and the ability for team members to focus collectively, both *inwardly* and *outwardly*, on a broad range of topics. Participants examine stakeholder issues, values, functions, cost, benefit and relative worth, with a view to building consensus on the best way forward. This significantly reduces subsequent project development time and identifies the optimum choice of strategy and components. Rigorous application of the process with the right personnel is crucial. Depending on project size, complexity and schedule, more than one workshop may be appropriate, and the study steps and pace can be adjusted to suit a specific situation.

Analysis by function and functional dependencies is very powerful. It is the crux of a successful value study and differentiates the value-based approach from other management philosophies and tools.

## *Risk management*

Through its logical process and multi-stakeholder representation, a value and risk management workshop is an ideal forum to identify and categorise project risks and uncertainties. Assessment of the risks may be made quantitatively, qualitatively or pragmatically. The synergy of the team invariably leads to an innovative and practical risk management plan.

Risk and uncertainty are present in quite different forms for various levels of project authority and implementation personnel. Risk may be stated as any potentially detrimental occurrence to a project in terms of cost, schedule, safety, quality, reliability, stakeholder disruption, etc. At a higher level, risk may be related to, say, failure to protect public health or economic failure of a region.

While neglect to identify a risk can be expensive, so can unnecessary allowance to avoid each and every possible risk that may be foreseen. Further, once identified, risk may be examined creatively and turned into a scheduling or economic opportunity. A balanced approach is advocated to specifying methods to evaluate risks. In line with varying stakeholder values, the degree of risk accorded to a project party is dependent on the level of comfort for carrying that risk. Risk transfer is a common practice, but may incur significant unnecessary cost if not properly evaluated.

Risk evaluation is seen increasingly as an essential part of smart project management. During the project development process, a particular risk can be a lack of stakeholder agreement on project needs and proposals. This can derail the approval process and incur significant delays in project development and subsequent implementation. Risks may be present due to limited experience, lack of information and general uncertainty regarding future conditions and viewpoints. Risks may also occur as parties, personnel and relationships change during the course of a long duration project. Construction phase risks often relate to costly unknown conditions such as poor ground, old buried structures, archaeological remains, future exchange rates, labour disputes, etc. During the project operational lifetime there may be significant risk of the project not performing as required (e.g. through technical failure or inaccurate data).

The value and risk management study applies the appropriate techniques and level of analytical effort within the workshop to proceed through to formulation of recommendations. It may well recommend a more thorough risk evaluation to take place separately.

## Value and risk management application stages

The value methodology is used at various stages of programme and project development for building consensus on situations and available options. It is applicable in most sectors. It is most successful when used as the basis of a long-term, continuous approach for business and project success.

The best results are obtained through early application. Ideally, the value process is used as early as the needs identification stage and for facilitating strategic partnering workshops, including development of dispute avoidance procedures. As the programme develops, the methodology is applied as a continuous improvement mechanism to ensure maximum cost effectiveness, functionality and appropriate quality. It can be applied with surprisingly good results as late as the tender assessment and contract negotiation stage for traditionally developed projects. Accordingly, the focus changes over time.

The value approach is applied best as a pre-ordained, step-by-step series of events, carried out at key points. In this way, the value and risk management program sessions may be better planned and integrated to become invaluable milestone mechanisms for continuous improvement and to address changing project and procurement conditions. This results in better understanding of the context in which issues are judged, through the team members seeing the issues in the same light. Consequently, better commitment through 'ownership' of decisions is established and 're-visiting' of project decisions, with related rework, is avoided.

It is noteworthy that the process of building consensus is not linear, as tends to be the case for traditional project management. Consensus building requires successive iterations to review and confirm or modify various interim decisions. This actually speeds-up the overall project development schedule as it pre-empts many otherwise time-consuming meetings outside the workshop(s) process.

## Summary

Various authoritative sources have advised the need to address more rigorously the client's business case. The value and risk management approach may be applied to a wide range of small, complex projects through to large projects or programmes. Early application of the value methodology as an integral component of the strategic procurement process leads to significant savings relating to schedule, staff time, capital costs and lifecycle costs. In addition, experience shows that 'it is never too late' to derive substantial benefits in terms of enhanced functionality, team building and cost improvement.

A comprehensive value and risk management programme will leave behind, for future reference, a set of milestone documents recording how strategic direction was set and the basis of key decision-making over time, as project circumstances, stakeholders and development personnel inevitably change.

## 7.3 Soft value management

Contributed by Professor Stuart Green, University of Reading, United Kingdom. Stuart Green has long been an advocate of value management and continues to argue that the concept must not be allowed to stagnate. He frequently challenges taken-for-granted assumptions and through a series of funded research projects has sought to strengthen the theoretical underpinnings of value management (VM). He believes that there is a direct connection between the way that VM is conceptualised and the way that it is enacted. Continued theoretical development is therefore essential to the long-term sustainability of VM practice. In this paper, Stuart presents his vision of 'soft' VM as an alternative basis for reflective practice.

The concept of Soft VM has been accepted by many for several years. Any explanation of Soft VM must start with an acceptance that organisations – and organisational theory – have changed beyond recognition since Larry Miles first articulated the principles of value engineering. Organisational theory in the 1940s and 1950s tended to conceptualise organisations as if they were machines pursuing pre-determined objectives. From this perspective, value-for-money is all about efficiency. The dominant question therefore becomes 'how can we achieve required function at least cost?' This is the principal model of value-for-money that continues to lie behind many conceptualisations of VM. Such recipes remain popular in the construction industry, where the allegiance to machine thinking has been continually reinforced by a series of industry-improvement initiatives.

A prime example was provided by the Egan Report in 1998, which advocated the adoption of 'lean thinking' for the elimination of waste. Egan's definition of value management was especially telling:

> Value management is a structured method of eliminating waste from the brief before binding commitments are made. (DETR, 1998)

The above quote illustrates the recurring tendency to reduce everything to this simplistic quest for waste elimination. There is seemingly little understanding that value-for-money is initially dependent upon a robust process of project definition. In situations where the client organisation comprises multiple interest groups, this is not something that can be achieved through the semi-automated application of tools and techniques. It requires a different way of conceptualising client organisations; it requires a different way of conceptualising value-for-money; it requires a different way of *thinking*. The argument is that the ongoing obsession with waste elimination is hindering the achievement of value-for-money. Hence the need for Soft VM.

The first step towards understanding Soft VM is to recognise the limitations of the established lexicon of waste elimination. There are numerous valid alternative conceptualisations of value-for-money that are too often squeezed out of the conversation. The essential starting point is to understand the managerial discourse favoured by the client organisation. It may be that their overriding aim is to retain flexibility to adapt to future scenarios. From this perspective, value-for-money must be linked to the inherent flexibility of the proposed solution. Any attempt to collapse the debate into a singular interpretation of the 'required function' will miss the point – as Sir John Egan obviously did.

Other clients may place great emphasis on knowledge management, in which case VM would be better conceptualised as a learning exercise. Another common possibility is that the client is multi-faceted, with different interest groups possessing different interpretations of where their business should be heading. This is by no means dysfunctional, and is perhaps an essential requirement for the modern organisation orientated towards change and innovation. In such a circumstance, the client is most usefully understood as a political entity; the purpose of VM is to build and maintain a political constituency across disparate stakeholder groups. Yet another possibility is that the client may see the new project primarily as the means to support a wider programme of culture change, in which case VM would be seen as the means of challenging prevailing assumptions.

Of course, none of these possibilities are new to value practitioners used to working with clients at the front end of projects. It is Soft VM that provides the necessary models to support these diverse interpretations of value-for-money. It must further be recognised that practitioners must be able to persuade such clients that they can make a positive contribution during the early stages of project definition.

Those whose storyline is limited to the tired discourse of waste elimination are unlikely to be persuasive. The poverty of the guiding conceptual model limits their involvement to the latter stages of projects (often in response to a projected cost overspend). Of course, Hard VM (otherwise labelled 'value engineering') has validity within such contexts. But the rewards and returns are much less than can be achieved through the strategic application of Soft VM during the early stages of projects. And here lies the rub. The practice of VM is inexorably shaped by the guiding theory. It is the underlying conceptualisation that shapes how practitioners sell their services, it shapes what they do, when they do it, and who they involve in the process. It also shapes the fees they charge.

*Soft VM draws from the body of knowledge known as 'group decision support'.* Group decision support is defined as 'any process that supports a group of people seeking individually to make sense of, and collectively act in, a situation in which they have power' (Bryant, 1993). Guiding concepts include a recognition that (1) decision-making situations are dynamic, ill-defined and multi-perspective; (2) people are conceptualised as active components; and (3) the 'problem' cannot be identified in isolation of the perceptions of human stakeholders. In essence, Soft VM is less about problem-solving, and more about problem-shaping. As such, it provides a radical departure from traditional value engineering.

The discourse of waste elimination and function analysis gives way to an alternative discourse of participation and consensus building. The umbrella term 'group decision support' includes a range of problem-structuring methodologies that can be tailored to meet the requirements of different situations. Examples include: strategic choice (Friend and Hickling, 1997), soft systems methodology (Checkland, 1999), robustness analysis (Rosenhead, 2001) and cognitive mapping (Eden and Ackermann, 1998). These are the core methodologies of Soft VM (see Rosenhead and Mingers, 2001 for accessible summaries).

It is important to emphasise that the above are methodologies in the broader sense, that is, they encompass guiding principles that must be adapted to suite particular circumstances. Each methodology is informed by a distinctive theory of organisational decision-making. Intelligent adaptation is welcomed. Soft VM combines intellectual rigor with an inherently practical orientation. Within the context of value management, the ideas were originally piloted through a series of research projects at the University of Reading during the 1990s. Stuart Green has since implemented them on numerous occasions in the early stages of major projects. Many of the firms that collaborated in the early research have also absorbed the ideas of Soft VM to enhance their competitive advantage.

*If value management is allowed to stagnate, it will perish.*

## References

Bryant, J. (1993) Supporting management teams, *OR Insight*, **6**(3), 19–27.

Checkland, P. (1999) *Soft Systems Methodology: A 30-Year Retrospective*, Wiley, Chichester.

Eden, C. and Ackermann, F. (1998) *Making Strategy: The Journey of Strategic Management*, Sage, London.

Friend, J. K. and Hickling, A. (1997) *Planning Under Pressure: the Strategic Choice Approach*, 2nd edn, Butterworth-Heineman, Oxford.

Rosenhead, J. (2001) Robustness analysis: keeping your options open. In: *Rational Analysis for a Problematic World Revisited*, 2nd edn (Eds J. Rosenhead and J. Mingers), Wiley, Chichester.

Rosenhead, J. and Mingers, J. (Eds) (2001) *Rational Analysis for a Problematic World Revisited*, 2nd edn, Wiley, Chichester.

## 7.4 Value management in manufacturing

Contributed by Ken Scott, chairman of the Institute of Value Management (2003–5). Mr Scott is a leading practitioner of value management as practiced in the manufacturing sector. This paper outlines the evolution of value management in this sector and contains reliable lessons for the development and construction industry.

Value management had its roots in the US manufacturing industry in the late 1940s in the form of value analysis. Since then it has evolved considerably, and the application of value management in manufacturing differs in a number of important respects from construction. In this paper we trace the evolution of value management from its initial technical design focus to one which is focused on giving the customer value for money today, and identify the significant differences between manufacturing and construction.

## False start

Promising beginnings, value analysis (VA) which focused on existing products and value engineering (VE) which focused on new products was developed initially by Lawrence Miles in General Electric. They proved to be highly effective and although GE sought to restrict knowledge of the approach, it moved into the public domain during the 1950s. It spread from the USA to Europe and Japan and was initially applied in the UK during the 1960s. Although it was very successful in the UK in reviewing product design, the primary focus was not on improving value but reducing cost. This partially explains the fact that the interest in VA/VE was not sustained. The failure to sustain interest in VA/VE can be traced to two primary causes. First of all the ethos of value methodologies – defining value from a stakeholder perspective, empowering multifunction teams to use function thinking to generate alternative solutions and secure commitment through consensual decision making to win–win proposals – did not fit well with a largely command and control management philosophy. A philosophy that ultimately foundered on a win–lose conflict mentality captured in the winter of discontent. The second primary cause was that while cost was an important element in competitiveness in the 1960s, it became increasingly apparent that cost alone was not the answer. Unfortunately, value methods tarred with the cost reduction brush were not well placed to deal with the issue of Japanese supremacy in product quality that became increasingly apparent during the 1970s.

## Learning from Japan

There is a critical irony in Europe's lack of interest in value methods and their greater focus on quality. While the West initially focused on quality as conformance to specification (the definition of quality promoted by Crosby), Japan was focusing both on manufacturing quality inspired by Deming and quality in design inspired by Miles. It is instructive to note that only three Western management thinkers have been honoured in Japan with the Emperor's Medal of Honour, they are Drucker, Deming and Miles. So the West, stung by Japan's explicit manufacturing capability embraced quality management and largely ignored value methodologies that were the less obvious part of the foundation of Japan's success.

It was not until after 1985, when the Massachusetts Institute of Technology (MIT) undertook the largest benchmarking exercise ever undertaken in order to understand the reasons behind the Japanese and Tiger Economies successes in the automotive industry that the importance of value methodologies became apparent. The outcome of that exercise was popularised in the book '*The Machine That Changed The World*' by Womak and James. They determined that the Japanese performance and cost advantages had accrued by establishing a design to cost approach. This was based on setting performance and cost targets and then working with supply chain partners to use VE, VA and Kaizen to optimise design and drive waste out of the manufacturing processes and achieve the performance and cost targets. The authors coined the phrase lean supply to describe this philosophy. They later developed this concept into 'lean thinking' which has subsequently been recommended for construction.

## Recent developments

The feedback from the MIT and other benchmarking activities has been the spur for a resurgence in the application of VM in manufacturing, particularly in the UK. This resurgence has had two distinct but related foci – the improvement of existing products and the development of new products, both with an emphasis on the contribution of the supply chain. Most organisations have initially focused on current products (VA), only to realise that the greatest benefit comes from applying VE at an early design change.

The cost structure of most companies in manufacturing has changed dramatically over the last 30 years. Internal costs, particularly direct labour, is no longer as significant. It has moved from centre stage to perhaps 15% of total cost, while bought out costs for components and services is around 60%+ of total costs. Suppliers and procurement has therefore moved centre stage. The criticality of suppliers means that short-term cost exploitation is progressively being replaced with longer-term collaborative working. It follows that most VM activity is not simply multifunctional but multiorganisational and this has implication for facilitating the process. The facilitator needs to focus not only on the current activity and bringing together diverse interests but to recognise that longer-term relationship development are also important. The soft benefits of VM are now therefore more highly valued.

The approach in the early design phase has been to use value methods to determine the value criteria used by stakeholders to judge whether a product is acceptable (value profiling), then to use comparative VA to set performance and cost targets (value planning) and then use VE to achieve the design to cost targets established in the value planning stage. This combined use of value profiling, comparative VA and VE has been described as VM. In manufacturing, with its very heavy reliance on outsourcing, suppliers are critically involved in VM. Unfortunately this definition of VM does not directly align either with construction practice or the European standard on VM.

Most organisations are driven to apply value methods because of pressing current problems and these are most often raised by the procurement function and therefore involve their supply chain. The approach to current products differs from products in the early design phase in that it is concerned not only with product and process design but with process management. Value methods in this context therefore include VA applied to both products and processes (manufacturing and business processes) and the application of Kaizen or process improvement techniques.

The application of VM approaches typically achieves cost savings opportunities of 5–45%, plus improvements in performance. The range of cost improvement opportunity is generally related to the experience curve. The further down the experience curve, the more difficult it is to achieve high levels of cost savings.

## Construction and manufacturing

There are significant differences between construction and manufacturing VM practice. These differences can be explained partly by history and partly by their different contexts. The most obvious difference is the time made available to VM activities. The construction practice of seeking to undertake a VM study in a day, is by manufacturing standards ludicrous. Many manufacturing workshops last 3 days but place a much heavier emphasis on team development, mutual understanding, the development of team consensed proposals and work to a far greater level of detail. They rarely integrate risk to the extent that is becoming more common in construction.

The practice in manufacturing of applying value methods at a detail level through the use of function analysis can be partly explained by volume. This is at best a partial explanation as the much higher item costs in construction often offset the volume effect. A more likely explanation is the differences in prevailing cultures between manufacturing and construction. Detail application takes time, time to gather data, time to analyse, time to reflect, time to be creative. During this time the rush to a solution and activity is deliberately suppressed. Cultures that value action and discount reflection will dismiss detail application. While the greatest benefits of a value approach in construction are at the higher levels of application, the rejection of a detailed approach means that opportunities for improving value are lost.

Perhaps the biggest differentiator between construction and manufacturing is the emphasis on the work to be undertaken by the team. This reflects a different set of values and as a consequence a greater willingness to take a longer-term perspective in manufacturing. In construction everything is done to make the workshop as efficient as possible, to cut to the chase and focus on outputs. Time to develop as a team, to explore issues, to break down barriers and develop long-term relationships are severely restricted. Many of the soft benefits of a value approach which underpin the longer-term creation of a value culture are heavily discounted.

## Lessons to learn

There are important lessons that both manufacturing and construction can learn from each other in the application of value methods. Manufacturing can learn from construction's focus on high level application, the integration of risk and the stronger emphasis on efficiency. Construction may benefit from greater value opportunities through more detailed application throughout the supply chain, and may achieve longer-term benefits with greater value placed on the soft benefits of VM.

## 7.5 Value management for public services

Contributed by Clive Bone MBA CEng FIMechE FIEE Hon FIVM. Clive Bone of consultants Bone & Robertson has a local authority, public sector and industrial background and specialises in UK public service value management. Here he explores the application of value management in public services and the constraints that exist in delivering it.

## *Statutory background*

It is arguable that public services only use value management when forced by government policy. This inertia is not unique to the United Kingdom. Theodore C Fowler makes much the same point in respect of US public services while Mckinsey's UK Managing Partner, Dominic Casserley, argues that even private sector managers in the United Kingdom lack the incentives and skills to improve.

Accordingly, value management has only been used in the context of the public sector where there has been clear pressure to improve, and then only grudgingly. It took the unpopular policy of putting services out to tender in local government and the NHS (compulsory competitive tendering) to cause some public services to use one-off value analysis exercises to improve the efficiency of their service delivery. This enabled in-house services to win their bids to continue to deliver the service. Whenever value analysis was used the success rate in terms of winning tenders was 100%.

The incoming Labour government of 1997 introduced a best value policy in place of competition. The policy applies to councils, police and fire services, and to housing associations. The underlying principles of the best value policy are demonstrably the same as value managements. However, no provision was made for training in the methods and skills needed for a sound value programme, and so the policy failed to prevent high council tax increases. Nevertheless, there have been a few authorities that successfully used value management to address specific reviews and these have added to the stock of experience.

## *Terms, principles and approaches*

The terminology used in public service value management differs little from that used elsewhere. Value analysis is used here to describe the review of a single existing service while value management describes a programme of such reviews using value analysis and other methods. Public services, however, tend to use the term 'review' rather than 'study', while 'workshop' is used to describe key team activities within the review – the creative or brainstorming stage, for example. The three-day workshop of industry where several stages of the Value Analysis Job Plan are progressed is generally not used in public services.

The underlying principles of value management for public services are no different to those that apply to other sectors. There is the same need for clarity of objectives at the outset, for selecting the right team, for ensuring that customer and stakeholder needs are fully understood, and for the effective management of the review process and workshop facilitation. All the same principles apply, however, practice, differs a little in detail.

At the outset the team must clarify what the service is for and re-examine the objectives of the value management study in the light of this understanding. This is done using FAST diagramming methods (see Chapter 9). In practice the typical FAST diagram for a public service can have from 40 to over 100 functions in 4–5 tiers including the prime function(s). The FAST diagram has proven invaluable for public service reviews. As a result, its usage has spread outside of value management reviews to become a widely used analytical tool in general use – as in benchmarking, for example.

## *Using function analysis to clarify purpose*

Some years ago Cheltenham Borough Council received awareness-raising training in value management. The topic picked to illustrate function analysis was the tourist information service. The exercise revealed that its basic function was not, as originally thought, to 'inform tourists' but was actually to 'promote tourism' – a more proactive activity altogether. The distinction is important because the basic function shapes thinking about the services – after all, if the team are not clear about what the service is for, how can they improve it?

The use of function analysis over the years has produced some interesting changes in the understanding of service basic functions:

- Supported housing funded by a probation service was not there simply to 'house offenders' but to 'reduce crime'.

- Part of the role of traffic police was not to 'reduce speed' but 'control speed'.

- After much debate the prime function of education for 14–19 year olds was determined to be to 'enhance life chances'.

■ The purpose of the custody of a person arrested who was suspected of an offence was determined as 'investigate crime'.

## Transforming understanding in a hospital pharmacy

The purpose of the pharmacy department of a large teaching hospital was determined to be 'manage medication' and not 'dispense drugs' as initially thought.

This revelation caused a value management review to be widened from the activities of the pharmacy department alone, to explore how best to use pharmacy in the hospital as a whole. While the costs of the pharmacy department as such was in the order £2m the drugs bill was a further £8m, nurse time in dispensing drugs £5m and the cost of adverse drug reaction a further £5m – due to extending the patient's time in hospital.

By focusing pharmacy effort to advising doctors on the wards and by having technicians make up doses whenever possible in the lab, the rate of drugs bill increase was slowed, the incidence of adverse drug reaction checked and nurse time eased.

Thus what began as a review of the pharmacy, initially expecting to save up to 10% of £2m, or £200 000, ended up identifying savings of more than £2m – equal to the entire cost of the pharmacy service.

## The role of function analysis to balance of service provision

A police service was using function analysis to explore crime intelligence. The function analysis diagram highlighted in stark fashion that practically all the resources were devoted to the gathering and distribution of crime intelligence and very little to its analysis. As a consequence the police service in question recruited more crime analysts – usually support staff who collate crime data, check trends and who ensure that crime information with respect to suspects is joined up.

## Public service reviews

A typical large-scale service review using value analysis is likely to take something in the order of 3–4 months from the planning stage to the point where the team has proposals to report. When first confronted with this rule-of-thumb times scale, experienced value practitioners from industry and construction regarded this as too long. On the other hand, public service staff used to 'best value' regarded it as far too short because best value reviews can last for a year or more. While the need for widespread stakeholder consultation can take time, 3–4 months for a major service review seems to work out about right for a focused team fully trained in value management.

In public sector value management, it is necessary to set tough targets. Under the best value regime, however, people became used to not setting targets in terms of productivity, quality or savings – a trend reinforced by lax inspection. Indeed, some express surprise at the need to set tough targets, citing best value reports that contained not a single mention of money. There are dangers in not setting targets. Best value teams can, and have, used the methods and tools of value management to shorten and sharpen best value reviews without actually delivering real gains or savings. However, where public services have applied value management with rigour, savings of 15% or more of revenue, coupled with quality improvements, have often been achieved.

## The value analysis job plan and the '4Cs'

Under best value councils and other best value authorities are expected to address the '4Cs' – Comparison, Competition, Consult and Challenge. These factors are well known in value management – indeed, Lawrence Miles advocated comparison. Make or buy decisions and competition are not new either. Nor is the need to consult users and stakeholders and to challenge the status quo. We explore here how these are addressed through the value analysis job plan.

*Planning stage.* At the planning stage the team is selected for the review, the timetable determined and so on. With public services the study leader will need to ensure that challenging targets are established. *Consultation* and benchmarking (*comparison*) helps reduce resistance to setting targets.

*Information stage.* Wide *consultation* and *comparisons* usually take place during this stage where information is gathered about the service. This stage may take 6–8 weeks.

*Analysis stage.* Typically, this would comprise a half-day or 1-day workshop. The purpose of the service is *challenged* using function analysis methods that also delineate the function – cost structure of the service.

*Creative stage.* This takes place a few days later and comprises a second workshop of usually no more than half a day. Here the need for the service is *challenged* and alternative forms of provision proposed, and it is here that *competition* is included in the list of things to be evaluated. At the end of this session a master list of ideas to be evaluated is produced and follow-up champions identified.

*Evaluation stage.* The proposals selected during the previous stage are developed and validated. This is where the make or buy issue (*competition*) is explored further. This stage may involve further *comparison* and may result in *competitive* bids for the delivery of some services. This process may take 6–8 weeks.

*Report stage.* The Study Leader then prepares a report for top management approval. The team, or its leading members, will be *challenged* to justify its findings. In local authorities this is often done by elected members (councillors).

*Review.* In accordance with good practice, there should be a follow-up audit in due course to ensure that the agreed changes have been made and the expected benefits achieved. In any event, authorities have specific duties under best value in terms of reporting back.

## Managing value management

Public services are good at setting up the structures of a sound value programme. They generally accord with the value management framework of BS EN 12973 save for the elements related to culture, training and methods and tools. These omissions invariably result in a lack of practical skills to make the machinery work. The resultant poor outcomes of the best value policy is primarily due to the lack of training of authority staff and inspectors – most are unaware that recognised training even exists for regimes of this nature.

## Council taxes and the future

Inflation busting increases in council tax and the continual drive by the HM Treasury to cut costs could lead to the increased take up of value management in local government. Councils and other public bodies are soon to be required to improve performance under a policy devised for the Treasury by Sir Peter Gershon. Value management is as relevant to this policy as it is to best value. Experience with value management in a broad range of public services has demonstrated that it will easily deliver the level of productivity improvements demanded by the policy.

The biggest obstacle to the spread of value management, however, is the government itself which is reluctant to promote best practice performance topics for fear of appearing 'prescriptive'. The lack of funding for topics such as value management is a major constraint and promotional effort to date has been done on a shoestring. How long this position can hold given the widespread unawareness of modern performance practice in the public sector, coupled with tax hikes and falling pubic sector productivity, remains to be seen.

## References

Theodore C. Fowler (1990) *Value Analysis in Design*, Van Nostrand Reinhold.
Casserley, D. Article in *The Times*, 22 March 2004.

## 7.6 Risk and enterprise

Contributed by Gerald Orman. Gerald Orman is a member of the Risk and Management of Projects (RAMP) committee of the Institution of Civil Engineers and an experienced and practioner of risk analysis and management in major projects. The following is a resumé from the July 2003 RAMP Conference that he convened at the Institution of Civil Engineers in July 2003, notes compiled by Luke Watts.

## Introduction

Introducing risk management into a corporate environment raises the question as to how focussing with the downside can be compatible with the entrepreneurial spirit on which private companies depend. The intention of the seminar was to look at risk management from the project concept through to the site hut and on to the boardroom. On the way, answers to some fundamental questions about risk management in a corporate environment for projects in an enterprise context would be sought. The conference considered where the responsibilities lay and what skills were required to undertake risk management, and how the results of the process were used to assist decision-making, without undermining enterprise.

## The contractor/investor's approach to risk and the context of major PFI

### Ian Rylett, Managing Director Balfour Beatty Capital Projects

Balfour Beatty Capital Projects (BBCP) focus on assessing the validity of operating a PFI project. Projects are selected based on the strength of the business case. BBCP develop and build the facility before operating and managing it under the PFI agreement. The kind of projects undertaken include infrastructure, roads, hospitals and schools.

The PFI projects require the long-term view and the risks involved in that. Money is earned in the later years of the contract but the major risks are taken in the early years. These may affect the commissioning and operational phases. To undertake PFI you have to be a risk taker to make money. Therefore the risks must be understood.

The key risk drivers revolve around the financial environment. It is important to maximise the price while remaining competitive. Debt must be financed in the cheapest way. The constraints must be understood. There is no such thing as risk free borrowing. BBCP use cover ratios to measure robustness, a typical structure having a 90–10 cover ratio, which requires a significant transfer of risk out of the special purpose vehicle.

The BBCP package risks for management to allow them to make better judgement about the shareholder return and look to which risks can be contracted out. The understanding of risks increases as the scheme develops. As much of the associated risk will be kept in-house as possible. The less contracting out the better, as this reduces the amount of contract risks, which allows more debt to be raised and the contract to be competitively priced. The quality of the deliverable is becoming increasingly key and this can impact on the likely earnings.

A low risk project will have a lower risk premium. A high-risk project will make it difficult to optimise the debt differential and creates uncertainty in pricing the project correctly. It will generate a greater premium if the pricing is right.

The BBCP only use risk evaluation tools that translate into a number. But professional judgement is the key to being a good entrepreneur. The numbers in themselves do not say much. It is what is behind them that gives the true meaning. The results of any analysis must be validated. The projects life is long, twenty years or so. How do you evaluate this, especially when personnel will change several times throughout its life?

It is difficult to bring the PFI back on track if something goes wrong. The financing defines the extent of the liabilities and it is still important to manage that exposure. Reputation can often be the biggest risk area for the shareholders as it can significantly impact on the share price and your ability to win future tenders.

So in summary the project must be efficiently structured. The rate of return must be understood. Long-term active management of risks is essential by building on the experience gained in past projects and optimising the bid preparation process.

## *Analysing and managing the overall procurement risk for prospective major building development*

### Deborah Vogwell, Central Risk and Value Management Team at Davis Langdon

Deborah Vogwell based her presentation on phase 4 of the Eden Project, where Davis Langdon are the project managers, with additional responsibility for the cost analysis and risk management.

She said that the main objective of risk management is to maximise the benefits of completing the project in line with its objectives, not just to complete within cost, time and quality parameters. Therefore the process is undertaken early in project lifecycle so that the outcome can still influence the project.

The Fourth Phase of the Eden Project was complicated but nothing is technically unsolvable. A large amount of risk is in the people involved in the project so it is important to build consensus and understanding so that everyone moves forward together.

The Eden Project is a victim of its own success, receiving more visitors than expected. This detracts from the experience. The fourth phase will improve the facilities to cope with the numbers and improve the educational elements.

The process began by interviewing all key project parties, not just project team but also the specialists such as the artists, horticulturalists, to gain a full understanding of the issues involved. The information gathered, including experience from the first three phases, was distributed to the workshop team.

The workshop was informal, involving six tables of six people. During a series of short presentations, communicating the vision of the project, the team aligned their understanding of the project. They then reviewed the issues arising from the earlier phases and agreed actions to embed the lessons to be learnt. The team went on to build a detailed function model, aligning the value drivers to the triple bottom line of sustainability, economy, environment.

The value driver model informed the understanding of the risks to create the conditions for success. This requires asking the right questions using experience from similar projects. Shortfalls are defined and assessed to consider if they could be severe and management action to mitigate the risk identified.

Finally, the workshop set up a small risk management group, chaired by the project manager, to manage the risks on an ongoing basis.

## Analysing the construction risks in a major rail project

### Barry Bray, Union Railways, Channel Tunnel Rail Link

Risk management is the systematic process to remove impediments to project performance, and ensure real project success. The process is both qualitative and quantitative.

The qualitative process is used to consider the objectives of the project and put the focus on the key risks by ranking them. The potential severity of the risk is considered and action plans for reacting to the risk defined.

The quantitative process considers the feasibility of the set of risks in the risk register. The process is interactive and runs throughout the project design phase. Once it is put out to tender the contractors will define their risks. Post award, both registers will be reviewed and the facts laid out for all to consider. At this stage discussion will revolve around which risks will be transferred or shared, based on the party best qualified to manage the risk.

On an on-going basis the contractors' top risks will be reviewed each month. Only risks on the critical path will be managed through the process. At the end of project the residual risks will be considered and passed onto the operator.

Minimisation of the risk is the key to the process. The earlier the risk is identified the more chance there is to mitigate it.

If the risk has a 100% probability then it is not a risk. It is an issue that you have to get involved with immediately and manage.

Computer modelling is not an accurate process and the answers do need to be interrogated. If the model is not trusted then it should be revisited and interrogated further. This process helps focus attention on the key risk areas.

Risk is the responsibility of all parties. The project risk manager does not manage risks but manages the risk management process, maintaining the risk register, and communicating it to the project parties. Top management buy-in is required to put some weight behind the process and drive the team to use it. Risk management must be interesting to get people to manage their risks and not to ignore them.

## *Discussion following the first three speakers*

Governments do not formally consider the risks in the initial phases. Therefore, the decision is often taken to proceed and the finances committed without thorough consideration. The costs then go up as the project develops.

The last ten years has seen insurance arrangements become much tighter, thereby encouraging the management of risk. Insurance is there for a rainy day. An organisation should look after risk before falling back on insurance.

People do not transfer lessons learnt enough. You see projects where the same risks happen in phase 2 as they did in phase 1. There is a need to ask the right questions and maintain a database based on past projects. Benefit delivery is everything, and operational objectives must not be neglected.

The legal team is not usually involved until things go wrong (excepting drafting the contracts). They are not involved in the early processes. They tend to focus more on commercial and the transfer of risk.

The client can be the biggest risk. Getting everyone to understand the project is essential as each party has its own interest. The person seen as the client is only a representative. The workshop approach should include at least 50% client attendance. Spectacular cost increases can occur due to scope change. The key thing is to communicate clearly and focus on achieving success.

Allocation of risk is not always successful. Some cases get it right but there are some cases where not enough is transferred. Making sure who is best placed to own a risk is a real challenge. For example, the risk of late access to a site. This is a risk that will impact the client but they may not be best placed to address it. Therefore, it is transferred to the contractor.

## The role of the in-house specialist

### John Fox, BP Exploration and Operation

The focus of risk management is to minimise unacceptable risks. For example, 30% of the costs of the Columbian business cover the risks of operating in such a country and the money is primarily used for employee protection.

To avoid major disasters BP have a capital governance process with finance memorandums and a risk profile identified early in the process. Each individual project has a control structure that includes 'stage gates' where the risks are communicated to a specialist team. The 'stage gatekeeper' can kill a project at this stage.

All stages require technical presidents to approve the move into the next stage. It is their responsibility to ensure that all risks have been identified, including opportunities. Some flexibility exists around how much risk they are willing to take, as long as there are adequate routes to mitigation. Where the risk levels are higher, the project will have to accept that they will need to meet stiffer hurdles on cost.

To support the process there is a risk forum made up of all people involved, and a risk and uncertainty guidance system. Most risk analysis is undertaken by individuals who operate independently from the project but use input from the project.

In one $4 billion project to lay a pipeline at a depth of 1.5 miles, new technology and innovation was used to reduce the project from 7 to 2.5 years. The functional specification was given to the supplier of the sub-sea system but they forgot to measure the hydrostatic

> pressure. This meant that seals, costing a mere 50p each, were omitted, leading to costs of several million dollars. The risk management process is intended to help reduce this type of incident.

Most organisations are poor at learning from past mistakes. BP have a number of risk systems, and Risk 2000 is a 'reservoir' used to address this weakness by providing a central repository for risk.

The role of the risk analyst is to facilitate and support the project but is it the leader who is responsible for risks. Where most go wrong is not in learning the lessons from the past and transferring knowledge effectively.

## Analysing the investment risk for bondholders in modern infrastructure projects

### Sandra Pereira, Business Development Group, Standards and Poor's

Standards and Poor's rate loans as well as bonds. The rating grades run from AAA to D. The cut-off point for the investment grade is BBB. Basis points vary and the ranges between BB and BBB may be 300 points. A default is classed as 2 days late payment. Their ratings are used to assess the creditworthiness of organisations or projects and are vital for raising funds. Most project ratings relate to infrastructure projects such as rail and road.

The table in Figure 7.1 shows the cumulative average default rates for the different ratings.

## The responsible approach to enterprise risk for the modern contractor/investor

### Linda Duncan, Management Assurance Services, KPMG

Not that many companies manage risks well or even understand their risks. This goes beyond the few scandals that have been well publicised. There have been 150 distress signals from 90 companies. There has been a link between those that have a good risk approach and those that have implemented it as a 'nice to have'. This is demonstrated by some of the recent corporate failures.

The upside of corporate governance is that it helps improve profits and sustainability. McKinsey believe that the global premium to the

Cumulative average default rates

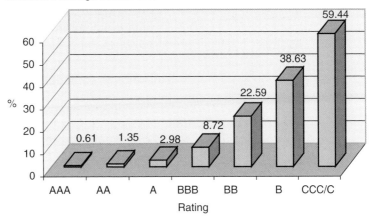

**Figure 7.1** Standard and Poor's cumulative average default rates (*Source:* Standard & Poor's Annual Global Corporate Default Study: Corporate Defaults Poised to Rise in 2005, January 2005, p. 17, Data: Year 15)

market price of a company is improved by as much as 12% if the company can demonstrate a good risk management process.

So what is corporate governance? It is a process to assist in the assessment of the companies' direction and to ensure the control of the company. The process specifies the rights and responsibilities of the board and other managers, while defining the procedures for making decisions. By doing this it provides a structure for setting business objectives. This is not dissimilar to governance within a project environment.

Risk management is a fundamental part of corporate governance, which allows the business to go to market and demonstrate they are worthy of trust in running the company. There is a lot written about the subject but there are very few good enterprise risk management books.

There are three lines of defence for an organisation. The first line is board responsibility. They must define a robust strategy that sets the future direction of the business. The culture and the people are the second. The culture must be right so that people are able to make the right decisions and to ensure that they are able to accept that the governance structure is necessary. The decisions people take will influence the achievement of the strategy. Performance monitoring is the third. It is important to ensure that the information is flowing up through the organisation on how the business is performing.

The use of project risk management and operational risk management is necessary so that there is a strong line of defence with underlying support. It is a matter of setting the right mindset. The business must understand what their role is in the risk management process and feed into it on a day-to-day basis. The risk management function is a roving role tasked with embedding the process into the business and encouraging that mindset.

An Enterprise Risk Management (ERM) framework should move from being reactive to being strategic. If something goes wrong you fix it. You must have a robust systematic process applied across the business to prevent it from happening again and again. This robust process will reassure top management.

The five stages of development of the ERM framework are:

1.  *Strategy*. If there is no policy in place then the process is likely to be reactive with no risk-based scenarios being considered.

2.  *Structure*. Forming a risk committee and embedding risk management into the organisational structure is essential. It will focus on the benefit at each level in the organisation. This enables the necessary contribution from each level and the ability to delegate responsibilities down the organisation.

3.  *Portfolio*. This is where you begin to take a considered and evaluated view across the business, projects and strategic direction.

4.  *Measuring*. This is the stage where you consider what will stop the organisation achieving its objectives and what will stop the achievement of growth. It does not need to be complicated and needs only qualitative techniques. The key benefit is in understanding what management action to take to mitigate the risk.

5.  *Optimising*. This is commonly the least well-developed area. This is where the organisation informs understanding of the strategic objectives by including risk and reward concepts in order to optimise the business case.

People must understand the value of the process and feel that it is real. A workshop can achieve this. Risk management should not be seen as an academic exercise, rather people should understand why they manage risks and why controls are in place. It is important to get through to the hearts and minds, and move risk management up the corporate agenda.

A number of public companies are implementing enterprise-wide risk management to address statutory. The key to a good risk management process is to consider the value it can add to the business. It is up to each company to decide what they require and implement it so it suits their business.

## *Summary*

There were many interesting and significant conclusions to be drawn from this conference, including the following:

1.   Whatever techniques are applied, ultimate responsibility for the risk stays with the line manager responsible.

2.   Risk is meaningless except in the context of objectives.

3.   There are a wide range of techniques for evaluating risk at different levels of a company and different stages of project development, done by project teams, in-house risk gateway departments and by outside agencies on behalf of lenders and insurers.

4.   However it is done, risk analysis and management is an intellectual challenge and not a 'tick in box' process.

5.   The keeping and study of risk records is essential to the development of expertise and the corporate learning process.

6.   The value of risk assessment as an analytical technique can be seen from the success that Standard and Poor's have in allocating bond ratings to project startups (see Figure 7.1).

## 7.7  The impact of partnering on value and risk

Contributed by Mike Thomas, Mike Thomas Ltd. Mike Thomas became involved in new approaches to collaborative working when employed at Whitbread Plc. He is now a leading consultant in promoting partnering and other collaborative techniques to public and private enterprises.

## *Background*

In his preface to *Management Teams – Why They Succeed or Fail* Dr Meredith Belbin asserts that 'The lone helmsman, whatever his ability, is prone to mistakes and oversights which reflect the limitations of his knowledge and experience. The management team has become the stable alternative ... so long as the right combination of people can be found'. Belbin's findings were that we rely not only on each other's technical skills but also on our balance of team role profiles to function effectively and deliver what is required for us as a team.

However, finding 'the right combination of people' can be difficult to achieve in construction project teams which often comprise a collection of individuals pulled together at relatively short notice. These teams (perhaps more accurately described as groups) are frequently assembled without regard to their disparate corporate and personal goals (values) or to the individuals' personal team working or interpersonal abilities.

Effective team working within the construction project teams is hampered by the linear communication process which constructs a hierarchy of roles in which those at the extreme ends of the chain rarely meet or achieve direct communication, one with the other. This discontinuity ensures, for example, that the client does not communicate their needs or expectations directly to the tradesperson (except through the designer, QS and constructor, etc.) and the tradesperson cannot suggest alternative solutions directly to the client (except through the reverse route) regardless of the amount of value that may be added by their proposal.

The industry shorthand for this communications structure is 'supply chain' and carries all the connotations of 'weakest links' both in the individuals comprising the links and in the interfaces. Communications may suffer distortion after passing through a number of individuals and interfaces. This is demonstrated in the apocryphal First World War example in which a message from the troops 'Send reinforcements – we're going to advance' was read at headquarters as 'Send three and four pence we're going to a dance'.

Such a disintegrated approach to the management of projects and project teams carries a high risk of error through misunderstanding or a lack of knowledge of what the other members of the team are doing. It delivers lower value to all members of the team because they are concentrating solely on their own roles and selfish aims without considering whether or how this impacts on others.

Covey identified that 'if I focus on my own Win and don't even consider your point of view there's no basis for any kind of productive relationship. In the long run if it isn't a win for both of us, we both lose. That's why Win–Win is the only real alternative in interdependent realities', and construction project teams actually comprise many interdependent realities.

## The partnering approach

The industry's focus in recent years (since the reports from Latham in 1994 and Egan in 1998) has been on reaping the rewards that can be delivered by collaborative working by integrated teams, through the practices of a collaborative win–win approach, a process for which the industry has adopted the shorthand term 'partnering'.

The term 'partnering' covers a wide variety of practices. The three key components of a successful partnering approach were identified by Bennett and Jayes as 'mutual objectives, an agreed method of problem resolution and an active search for continuous improvement'. Partnering, thus, requires team members to work closely together, initially identifying each partner's objectives (their selfish value criteria for the project) so that these can be understood by all and turned to mutual objectives for all partners.

Perceived (imaginary?) restrictions of public sector accountability and EU rules appear to limit the opportunity for strategic (i.e. longer-term) partnering to few clients. However, the benefits from project partnering (quantified by Bennett and Jayes as between 2% and 10% of project cost) can still be realised through an approach which overcomes the linear, chain-based approach of the industry and which develops an approach whereby construction project teams work collaboratively. To do this they will identify and set out their mutual objectives, agree methods of problem resolution and set out the processes by which they will achieve continuous improvement. This will deliver added value which will be quantified in appropriate terms, normally in the currency of the mutual objectives (e.g. £x reduced costs, y weeks speedier delivery, z% fewer faults/defects).

The key processes that bind a partnering team include a joint approach to the identification and management of value and risk. This approach is supported by the ethos of the partnering contract PPC2000 which, in Clause 1.3, requires partnering team members to 'work together and individually in the spirit of trust, fairness and mutual cooperation for the benefit of the project' and in Clause 3.1 where it states that 'The Partnering Team members shall work together and individually . . . to achieve transparent and collaborative exchange of information and to organise and integrate their activities as a collaborative team'.

## Why collaborate 'In the spirit of partnering'?

Why should construction professionals (who know all there is to know about construction) 'work together... for the benefit of the project'? Put simply, if they do not, and if all members of the team do not understand and support each others' value drivers, they will not be able to deliver the aims of all the project team members and will be unlikely to get support from their colleagues in return. While 'win–win' runs the risk of being an overused expression, it neatly encapsulates these principles.

For a construction client to receive a project that is in line with their expectations or for design or construction professionals to receive payment that is in line with their expectations (in terms of expenditure, timeliness and working expectation) then all of those expectations must be communicated clearly throughout the team, challenged where necessary to ensure understanding and delivered by the whole team.

If it appears to one member of the partnering team that they will fail to achieve their expectations at the end of the project, then their focus may drift from delivery for 'the benefit of the project...' to delivery for their own benefit. Unfortunately, this step will probably mean that they are not concentrating on delivering the expectations of others and this will start the vicious circle of blame, claim and drain of resources that leads to projects failing to deliver to anyone's expectations.

In order to emphasise the collaborative nature of the partnered project, PPC2000 requires the partnering team members to 'establish and demonstrate best value to the client...' and requires 'partnering team members with relevant expertise...' to review and submit proposals for how each 'relevant risk could or should be eliminated, reduced, insured, shared or apportioned...'. Thus, value is identified and risk managed by the partnering team.

## Practical examples

Working collaboratively appears to drive additional costs (team workshops, meetings, hiring venues and facilitators). So... how is it that saving these costs results in lower value and increased risk?

1. Failing to work collaboratively leads to assumptions and frequently results in the wasted cost of having to put things right later (more cost = lower value).

2. A tit-for-tat blame-seeking approach when things go wrong (as they will) results in additional costs of one-on-one marking (higher resource use = lower value).

3. If team members do not talk, the probability of double-accounting for risk allowances/contingencies is increased. (I hold a sum in case you do not cooperate and you hold a sum in case I do not... resulting in higher risk pots and less expenditure against the stated aims of the project.)

4. Structured team workshops actually replace a multitude of *ad hoc* poorly planned meetings in which decisions are not taken, actions misunderstood and many attendees fail to contribute (high probability of a risk of misunderstanding within the team).

5. Failure to hold regular team events (social or work-focused) prevents the informal exchange of tacit knowledge (the village well approach).

Three very clear examples of collaborative team working delivering higher value *and* lower risk to all partners are demonstrated below.

Whitbread Hotel Company took a strategic partnering approach to the conversion of 20-plus hotels to Marriott brands in 2000. While some of the organisations who took part in this programme were formal strategic partners (constructors and consultants), others were contracted suppliers who committed to collaborative working because they understood that this approach would enable each and every project to be delivered on time and to the client's requirements. This would demonstrate their ability to work within the team and identify their potential for further work. As a result of regular team workshops and reviews, the projects were delivered within the overall programme budget, to time and to agreed quality standards.

## *Large project example*

The Open University new library team took a project that had previously had difficulties meeting its pre-contract cost target in a traditional, non-collaborative, tender route and succeeded in delivering the project to budget and time while exceeding client and end-user expectations by working collaboratively. Commencing in

> January 2002 the full team of clients (including users, estates, maintenance and the full design and construction team) met no less than quarterly for partnering, value and risk management workshops culminating in three reviews over the 12 months after completion of the project. The project met its targets for cost, time and quality – exceeding the expectations of the client.

## *Small project example*

> The project partnering team that delivered the refurbishment of 20 flats at Lukely Court, Newport for the Isle of Wight Housing Association included substantial representation from the constructor's supply team and the tenants. There were 22 attendees at the initial partnering workshop in February 2000. Clear vaue criteria were set for the project in the form of a partnering charter of mutual objectives. Risk was managed through a robust process for resolution of issues. During the project the team met regularly both in site meetings and at quarterly continuous improvement workshops where successes and issues were elicited and noted. Tenants were so pleased with the end result that they held a party for the construction team.

## 7.8 Just how mature are your organisation's processes?

Based upon a paper entitled: 'Just how mature is your organisation at project management?' contributed by Alan Harpham, Chairman of the APM Group Ltd. Alan Harpham comes from an engineering background, constructing major oil and gas facilities before majoring in project management. He was director of MSc in Project Management at Cranfield University and former managing director of Nichols Associates. His work with the APM Group Ltd involves setting standards and accrediting qualifications in project management skills with the Office for Government Commerce (OGC) and other major organisations.

This article explores the development of project management and its maturity. It describes the development of a set of questionnaires of the OGC based upon the Carnegie Mellon maturity model and how these may be used to determine the maturity of an organisation's project management processes. Its relevance to value and risk management is at a similar approach, and may be used by an organisation to assess its maturity in the processes described in this book.

The project management maturity model (PMMM) describes the key practices related to each of the project management process areas in terms of what an organisation should be doing to establish and improve its ability to manage projects effectively and the organisational capability generally.

The model identifies five maturity levels, each of which focuses on a limited set of key process areas:

Level 1: Initial process – few defined processes, *ad hoc* practices, much depends on individual effort and charisma.
Level 2: Repeatable process – successes in similar applications can be repeated.
Level 3: Defined standard process – all projects delivered within standard processes.
Level 4: Managed process – quantitative measures of process effectiveness and product quality are collected, understood and acted upon.
Level 5: Optimised process – continuous process improvements throughout the organisation.

Each process description is summarised under the following six headings.

- Functional achievement/process goals

- Approach

- Deployment

- Review

- Perception

- Performance measures.

These headings provide a framework for describing what an organisation should be doing in generic terms to establish and embed each process.

The model can be used in several ways, for example:

- To understand key practices that are part of an effective organisational process to manage projects.

■ To identify the key practices that need to be embedded within the organisation to achieve the next level of maturity.

■ To understand the rationale behind the assessment questionnaire.

In describing the project-related activities within key process areas, the model recognises not only the project management activities being carried out at the individual project level, but also those activities within the organisation that provide focus and help sustain effort to build a project infrastructure of effective project approaches and management practices. In the absence of an organisation-wide project infrastructure, repeatable results, that is, success, depends entirely on the availability of specific individuals with a proven track record. This does not necessarily provide the basis for long-term project success and continuous improvement throughout the organisation.

## The concepts of matrix management

In the 1990s major organisations began to develop management by projects to bring about beneficial change which led to the concepts of programme and benefits management that are practised today.

As the importance of project management to businesses has grown, so the processes have evolved from being relatively simple to complex. To enable organisations to benchmark their performance in project management, a number of maturity models have evolved. This paper describes a process for assessing an organisation's maturity level which was developed by APM Group using the OGC's PMMM as the standard to assess against. Rather than simply be a passive measure of performance, the model is designed so that it may be used by suitably qualified people to advise the organisation on how to develop, grow and mature further.

An organisation that is judged immature in project management terms may deliver individual projects that occasionally produce excellent results. However, the managers at both organisation and project level are more likely to be working in a reactive mode, that is, focusing on solving immediate issues. Project schedules and budgets are likely to be exceeded because of a lack of sound estimating techniques. If deadlines are imposed, project deliverable quality is likely to be compromised to meet the schedule.

A mature organisation has an organisation-wide ability for managing projects based on standard, defined project management processes that can be tailored to meet the specific needs of individual projects. The project approach is communicated effectively to project team members and stakeholders, and project activities are carried out in accordance with the project plan and the defined process. Managers are also capable of monitoring the progress of the project against the project plan, including the quality of project deliverables and customer satisfaction. There should be an objective, quantitative basis for judging quality of project deliverables and analysing problems with project deliverables, the project approach or other issues.

One would expect level 4 and 5 companies to be better able to manage projects with higher degrees of risk and be more likely to deliver service product offerings with more predictable levels of quality. Furthermore, organisations that can demonstrate a higher degree of process capability are more likely to be able to handle projects requiring higher levels of innovation. Projects necessitating higher levels of innovation will attract more contingency planning and budget to deal with the higher degree of uncertainty of outcome.

## *The assessment process*

To facilitate the process determining an organisation's maturity level, the APM Group Ltd. has developed a set of questionnaires for use by Accredited Programme and Project Management Consultants who will use the questions set out in the methodology and will determine the maturity level of the organisation. The organisation is recommended by the registered consultant for acceptance at their determined maturity level. An independent then reviews this maturity level and its determination and if in agreement accepts the organisation for certification at that level. The APM Group Ltd. carries out random audits of the work done by its registered consultants on the maturity model. To make the questionnaires as broad as possible in application they were also compared with other project management standards, such as the recognised bodies of knowledge, to ensure they were comprehensive in nature and to assist consultants in framing the recommendations for improvement. Registered consultants are licensed to use the assessment process through an accredited consulting organisation.

There are two stages to the process that can be used either by an organisation that owns its own projects or by an organisation that provides project management services to its clients.

## Stage 1

The purpose of Stage 1 is to identify roughly at what level of maturity in project management the organisation is. A Stage 1 questionnaire can be administered by the registered consultant, or self-administered by the organisation itself. The stage is based on 13 questions that should be answered as honestly as possible. The potential maturity level of the organisation is found by ensuring that the organisation scores an 'Yes' for every question at that level. This suggests the highest level the organisation might achieve in the more thorough evaluation in Stage 2 – a much more detailed set of questions. It is Stage 2 that derives the actual maturity level of the organisation and not Stage 1.

## Stage 2

The Stage 2 questionnaire is administered by a Programme and Project Management Registered Consultant (PPMRC) in accordance with the APM Group's accreditation process. To gain formal accreditation a full review must be carried out by an external and independent accredited consultant.

The PPM consultant will ask the organisational questions of the most senior person responsible for projects or project management and its improvement in the applying organisation. He will, wherever possible, verify the answers by reference to those interviewed for project questions or other senior staff of the organisation.

# 8 Study types

## Summary

■   This chapter details the main study types to which reference has been made earlier in this book. It aligns them with the project stage at which they are most likely to be undertaken and describes the main steps that the study leader should follow.

■   Study types that are covered include:

Value management study types

o   Need Verification

o   Project Definition

o   Brief Development

o   Value Engineering

o   Design and Cost Review

Risk management study types

o   Strategic Studies

o   Formal Reviews

o   Informal Reviews

■   The chapter also describes ways of conducting reviews to assess and improve project performance and learn lessons for continuous improvement.

■   It goes on to discuss the project cycle and how an operational review of a facility that is in use can provide the trigger for a new project.

■   Finally, it discusses how teams can build on their experience and make continuous improvements by invoking the Kano principle.

## 8.1 Types of study

Chapter 4 introduced the principles addressed by value and risk management studies at different stages in a project life. This section describes these studies in more detail. While value and risk study types are described separately, the experienced study leader will, in fact, cover both topics at each of the milestones at which formal studies are recommended.

The descriptions of most of these are practical to apply and achieve results. The description of each of the techniques should, however, not be regarded as prescriptive, but as guidelines for best practice and common usage.

Figure 8.1 provides a key to the study types and their relationship with key project stages (duplicate of Figure 4.4 for ease of reference).

## 8.2 Value management studies

### *Need verification*

The need for a project maybe triggered by several factors, these may include:

■ A change in the law with an impact on the business being undertaken

■ The introduction of a new service or product

■ A requirement for increased or reduced capacity

■ A change in the way a business operates or a service is delivered

■ An opportunity to exploit a new market opportunity

■ A desire to use a project to bring about beneficial business change

■ The acquisition of a new business

■ The outcome of an in-service review

■ Other reasons.

There may be several different ways in which a business may achieve the benefits sought arising from any one of the above sources. Not all of these will involve constructing a building. It is

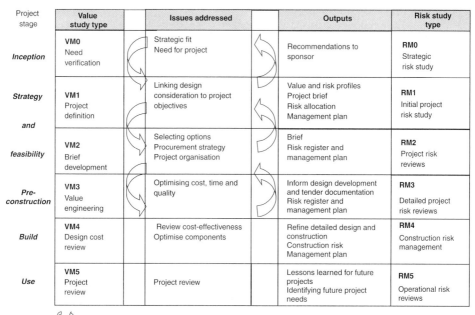

| Project stage | Value study type | | Issues addressed | | Outputs | Risk study type |
|---|---|---|---|---|---|---|
| *Inception* | **VM0** Need verification | | Strategic fit Need for project | | Recommendations to sponsor | **RM0** Strategic risk study |
| *Strategy* | **VM1** Project definition | | Linking design consideration to project objectives | | Value and risk profiles Project brief Risk allocation Management plan | **RM1** Initial project risk study |
| *and* | | | | | | |
| *feasibility* | **VM2** Brief development | | Selecting options Procurement strategy Project organisation | | Brief Risk register and management plan | **RM2** Project risk reviews |
| *Pre-construction* | **VM3** Value engineering | | Optimising cost, time and quality | | Inform design development and tender documentation Risk register and management plan | **RM3** Detailed project risk reviews |
| *Build* | **VM4** Design cost review | | Review cost-effectiveness Optimise components | | Refine detailed design and construction Construction risk Management plan | **RM4** Construction risk management |
| *Use* | **VM5** Project review | | Project review | | Lessons learned for future projects Identifying future project needs | **RM5** Operational risk reviews |

Arrows thus indicate potential reiterations which may be necessary if circumstances require strategic changes to the project

**Figure 8.1** Relationship of value and risk studies with project stages

the management's task to identify the most beneficial way for the business to achieve the desired benefits.

Thus the first stage of any project is to verify the need, if any, for a building project. A study at this stage will include looking at whether the proposed project fits strategically with the direction in which the business is going. It must identify the needs of the business, of each of the users and of other stakeholders who may have an interest in the project. It must clearly identify the benefits that are sought. It must develop an understanding of why the project is being considered now and what has changed to make it necessary. It needs to define the required strategic functional performance of the project in terms which can be easily conveyed to people who are not closely connected to the project or the business. And it needs to explore all possible options for delivering the benefits, not simply the solutions involving buildings.

The output from the study will comprise recommendations to the project's sponsor. The purpose of these is to help the project's sponsor to make a decision as to whether or not to proceed with a project and if so, which project. At this early stage there is no one-way to undertake the above tasks. The case study further illustrated how these objectives may be achieved.

An international business had identified a large number of projects, some large and some smaller which would improve its business operations. The total estimated cost of undertaking all the projects exceeded, by a significant margin, the available budget. It was therefore necessary to select only those projects that were afford-able and added most value. The study began with an anchor workshop at which the team agreed on the principal benefits to be derived from the programme of projects. They then identified the key value drivers (or primary functions) to which each of the projects should contribute. The anchor workshop also identified the key risks pertaining to the programme. A period of several months' consultation with the stakeholders (primarily the user groups) involved with each of the projects followed.

This information was used to build an outline business case for each stating:

- the project objective

- why it was needed

- who supported it

- a SWOT analysis

- a financial analysis to calculate the net present value (NPV) and internal rate of return (IRR) for the project

- the timescale and

- the overall capital budget.

Each of these outline business cases was then validated with the stakeholders and tested for strategic fit with the overall value driver model identified at the anchor workshop.

Validated documentation was then sent to each of the stake-holders and a series of workshops were convened attended by the appropriate stakeholders. The objective of each workshop was to place each of the projects into one of three categories, namely: (1) recommendation to proceed with development of the project; (2) recommendation to put on hold or revisit some of the assumptions made in developing the outline business case; and (3) recommend that the project be abandoned. Each of the work-shops, because of the amount of preparation that had been done beforehand, was able to review several projects. Techniques such as the combinex option selection technique (see Chapter 9) were used to arrive at consensus decisions. The main evaluation criteria included the programme level value drivers and risks. Finally, at a wrap up meeting, the recommendations were presented to senior management to enable them to decide which projects should go forward.

## *Project definition*

Having decided to proceed with a construction project it is necessary to clearly define the project in terms of the benefits to the business and to do this in a language which conveys the project imperatives to the project team. It is likely that the project definition study will take place at the beginning of the concept development stage. Studies at this early stage in the project are hugely beneficial to the client body and project teams alike, since they provide valuable opportunities to learn from each other's experience and reflect upon business needs and wants.

On the client's side, the process will force them to express their needs clearly in unambiguous terms which avoids much misunderstanding later on. From the project team's point of view, who will be new to the project, even if they have worked together on other similar projects with the same client, it presents an opportunity to interrogate and challenge the client's requirements in a non-confrontational way and thus gain a far better understanding of their underlying needs. There are two types of studies commonly used at this point.

### Stakeholder conferencing

The first of these is stakeholder conferencing. In the early stages of projects which involve large number of stakeholders, for example, regeneration projects, it is essential that the interests of all stakeholders be considered, if the project is to be a success. This requires working with large groups of people. The ideal group size for most of the value and risk study types described elsewhere in this chapter is relatively small (less than, say 20 people). To involve large groups of, say, 50 people or more, a different approach is needed.

The purpose of the study is to build a shared understanding of the project objectives and aims between the sponsoring client body (possibly comprising several organisations) and the stakeholders. It also provides an opportunity to reconcile the stated needs of each of the stakeholders and explore trade-offs to optimise the outcome for all.

*Strategic briefing meeting*

The study will normally start with a strategic briefing meeting with the client body and some of the key stakeholders. The study leader will chair the briefing meeting. The agenda will include:

■ A discussion around the underlying business needs.

■ Alternative solutions that have been considered, some of which may still be valid options.

■ Identification of some of the major strategic risks (mainly of a business nature) to the project.

■ The underlying assumptions to the outlined business case and validate that these are still relevant.

■ How the job will be financed and address affordability issues.

■ The format and timetable for consulting with the stakeholders and the convening of a stakeholder conferencing workshop.

■ How people will work together constructively.

The value and risk management team will then interview stakeholders in groups or individually and gain from them a good understanding of the issues that they wish to address.

They will analyse and collate the information into a workshop handbook which will set out the following:

■ The time, place and location of the workshop.

■ The purpose of the workshop.

■ The scope of the ground to be covered during the workshop, including items which are outside the scope of the workshop.

■ The expected outputs from the workshop.

■ Preparatory work to be undertaken, if any, by those attending the workshop.

■ A list of the documents which should be familiar to those who are participating in the workshop.

■ Copies of information that have been gathered by the facilitator team, analysed before the workshop.

■ An agenda.

*Workshop*

If the number of participants exceeds 20, the study leader is likely to need more than one facilitator to help him run the workshop. One facilitator between two groups of six people represents a good balance. Much will depend upon the skills of the facilitators, the nature of the project and the stakeholders who are represented. It is normal to hold the workshop in a single large room where the delegates are arranged in tables of 6–8 people per table (the author has found that it works better with 6 than with 8), with plenty of space to move between tables and flipcharts or other visual display units at each table. The study leader's team will have a separate table somewhere at the front of the room together with audiovisual equipment to display information. The workshop will alternate between plenary sessions and group working sessions as the teams work through the issues.

*Seating arrangements*

Seating arrangements are a matter between the study leader and the client, and will depend upon the nature of the study. There are two schools of thought. One is to let people where they will, being attracted by the subjects being addressed at the individual tables. This has the advantage that people address the subjects which interest them but suffers the disadvantage that one can get cliques developing, who may have their own agenda, which differs from that of the overall study. The other approach is to think out seating plans carefully to ensure balanced representation at each table (which may change during the day depending on the subjects being discussed). This has the advantage of delivering a more balanced outcome but suffers the disadvantage that some people may be disappointed with the subject matter to which they contribute and feel that they have not contributed in areas of prime interest to them. There is of course of no reason why, if people feel strongly enough, they should not move from their appointed seating arrangement.

The business of the day generally follows the following format:

*Introductions and confirmation of workshop expectations*

It is likely that many people will not know each other and therefore the facilitator team should provide labels stating their names, organisation and designation on the project. A dinner or similar social event for key contributors the evening before the workshop will help people to get to know each other in a relaxed environment.

This also provides the ideal opportunity for senior management in the client body to put across some of the key messages that they wish the workshop to address. Because of the number of people involved, individual introductions are seldom practical. Confirmation of workshop expectations may be achieved by asking all participants to state their expectations on flipcharts which are provided in the reception foyer to the room while coffee is being served on arrival. This also provides a 'buzzy' way for people to get to know one other and understand each other's concerns.

*Presentations.* The study leader will outline the format of the workshop so that everyone understands what is expected of them. He will invite key individuals to address the entire workshop to confirm what they wish to get out of it, summarise the information that has already been circulated and most importantly, impart any new information that may have come to light since. These presentations should last no more than about five minutes each and may be accompanied by visual aids. The recorder to the workshop should gather up copies of visual aids and handwritten notes for later incorporation into the report.

*Group work to discuss first topic.* The study leader will then describe the first topic for discussion at individual tables. Normally this would be an exercise to improve understanding of the purpose of the project. It could, for example, address a draft function analysis of the project, linking the vision, project objectives and key value drivers (or primary functions) and ask each table to develop different parts of this model. The facilitators will work with the teams, helping them to develop their ideas and record them on the visual aids provided. The workshop participants should record the data in their own words rather than have the facilitator do it for them. One reason for this is that this is quicker. Another is to ensure that the words used accurately reflect the group's feelings so that they can identify with them.

*Plenary feedback.* After the groups have worked on their appointed tasks for about half an hour or so, the study leader will ask them to reconvene into plenary session and give feedback to the whole group on what each table has developed. He will chair a short discussion during which comments are received on each group's outputs. The workshop will then break into groups again, either to further develop the same task on which they were previously engaged, incorporating the comments from the plenary session, or a new task.

*Other tasks.* The next task is likely to be aimed at building understanding of the main risks on the project, possibly from an initial risk listing that has been prepared beforehand. Each table will address different subjects under the initial risk listing and after half an hour or so, reconvene in plenary session and provide feedback on what they have developed. The workshop will continue in similar vein throughout the day alternating between group work and plenary sessions.

*Summary.* At the end of each day the study leader will summarise what has been achieved and explain the next steps in the process.

*Report.* The report will summarise the outcomes of the workshop and the consensus positions that have been achieved. It is likely to comprise two volumes, one slender volume containing a summary of the outcomes, intended for the senior management, the other a record of all the information that was gathered and discussed during the workshop.

A stakeholder conference is likely to generate a huge amount of information, the analysis of which can take a long time. It provides an ideal forum for all stakeholders to express their views, challenge the initial proposals, make trade-offs between their interests and develop consensus. For large projects, this will be far more effective than a small management team attempting to impose their wishes upon a reluctant public.

It will also lead to a much more rounded project. It addresses one of the key requirements of the best value programme for UK local government, that of *consultation* with the stakeholders and *challenging* proposals. It is a very effective way of genuinely engaging with the public.

A railway company wanted to explore a scheme to reinvigorate a substantial section of the railway network. It therefore tasked a small group in its internal projects department to draw up a brief for commissioning consultants to undertake a feasibility study. In order to test the appropriateness of the brief with the major stakeholders, the rail company commissioned a value management study. This included holding a 2-day stakeholder conference with around 50 senior managers present. These managers represented the train companies who would use the line, local authorities through which the network would pass, central government representatives, professional advisers, train operating companies and technicians. The purpose of the conference was to gain consensus for the feasibility study brief.

Delegates were placed at eight tables, each table comprising a suitable mix of expertise and interests. Four facilitators helped delegates to work their way through a structured programme of activities, each designed to identify and agree the key project deliverables and an implementation plan.

Participants were committed to the broad aims of the project and expressed a willingness to work together to achieve mutual benefit. While there was broad agreement, there were also important differences in emphasis. External stakeholders, principally train operators, placed almost twice as much importance on a cluster of value drivers comprising reliability, maintainability and adaptability than internal rail company stakeholders. By contrast, internal rail stakeholders were three times more likely to be focused on the importance of the business case.

A high level of consensus masked detailed differences that existed between all stakeholders. The key issues focused on the following headings:

- Delivery
- Capacity
- Performance
- Commercial
- Communications
- Integration
- Timing.

Detailed discussions highlighted a number of significant issues that must be overcome including:

- Lack of agreement on demand
- Potential national resource constraint to implementation
- Need to integrate the needs of rail services from outside the region served by the network
- Information for sizing facilities

- The need for a new maintainability strategy due to extended timetabling

- Overall journey time is more important than speed

- Provision of better located stations and depots

- Implementation strategy to balance new technology, major change with the need to minimise disruption

- Identification of non-user benefits to improve the business case, especially 'political' image.

The outcome of the study, which the overwhelming majority believed to be beneficial, included a significant redrafting of the feasibility study brief as well as a resolution to conduct similar consultation exercises on all future major projects.

## Conventional workshop

For situations where there is not the need to engage with large numbers of stakeholders, it is possible to convene a more conventional workshop with 10–20 participants. There is still a need for thorough preparation beforehand if best use is to be made of the workshop time. At this stage of the project it is likely to involve senior people who are an expensive resource and not given to debating trivial issues. The preparation therefore should distil the essence of the items to be addressed and analyse the information gathered so that the workshop can concentrate on processing it. A very effective agenda for such a project definition workshop runs as follows.

*Preparation*

The process starts with a strategic briefing meeting chaired by the study leader addressing the similar points to the stakeholder conference style workshop described above. The study leader will work with the members of the project team to gather and analyse the information ahead of the workshop, typically comprising:

- A description of the aims and objectives of the project, and an analysis of the key value drivers (or primary functions) linked to the considerations that the design team need to take into account in order to deliver them in full.

- An initial listing of the main risks to the project grouped under appropriate headings such as those proposed in Chapter 3 and assessed for likelihood and impact.

- Identification of the main stakeholders to the project, both internally and externally, analysed to show their interest in the project (positive or negative), and their influence, that is, a stakeholder map.

- Financial information including a summary of the underlying business case.

- Other background information as may be appropriate.

The study leader will assemble this information and distribute it to the workshop participants along with a formal invitation and documents describing the format of the workshop to which they are invited.

### The workshop

At this stage in the project it is likely that the workshop will be of short duration, half to one-day being common. The workshop will begin with introductions and confirmation of workshop expectations, in a similar manner to the stakeholder conference workshop. This might take place over an informal buffet breakfast.

### Presentations

Key stakeholders will state the main issues that they would like to address and share any new information with the project team. The study leader will record key points openly so that they can be seen by all present in the room. The list of issues and observations serves several purposes:

First, it provides a record of the key points that were raised.

Second, it provides the opportunity for members of the workshop team to question whether their understanding of the points is correct.

Third, and most importantly, it provides the basis of building better understanding between all participants.

Finally, it identifies a number of key issues which should be addressed at some stage later in the ensuing workshop, as illustrated in Table 8.1.

### Function analysis (or value profiling)

The study leader will table the draft value driver model (or value profile) if he developed one before the workshop, explain the logic

**Table 8.1** Issues and observations

| | Issues and observations to be addressed later in the workshop |
|---|---|
| 1 | Use only proven technology |
| 2 | Be aware of issues that might damage the brand |
| 3 | Taxation will be affected by the method of financing |
| 4 | A rigorous decision-making process will be required |
| 6 | Clarify the legal position on site acquisition |
| 7 | Need to balance demands for flexibility with redundancy of space |
| 8 | Review procurement programme to see if target end date is feasible |
| 9 | Seeking to create a larger pool of customers in which to fish |
| 10 | Must maximise GFA against planning constraints |

behind it and invite the team to develop it (or build it from scratch). This can be a very lively session. Discussions may range from identifying the value drivers to the precise wording that is used to describe the elements to accurately convey the quality of facility that is expected. The team may challenge whether certain features really are there for the reasons that the model may indicate. It is good practice to invite members of the workshop to write up their own contributions using sticky notes or similar means since this encourages interaction and debate. Sometimes this session will result in quite a different interpretation of the project from that which was originally tabled. There is nothing wrong with this since the whole project team has been on a very steep learning curve and the resultant model will reflect the outcome of their learning in an optimised project.

*Risk analysis*

The study leader tables the initial risk listing to validate that it provides a fair reflection of the risks that people identified during the preparation period. He then stimulates a high-level debate by asking questions directed at what appear to be weaknesses in the general project management infrastructure. Table 8.2 shows some typical questions that the study leader might ask under each of the risk categories.

*Note*: These examples in Table 8.2 are generic questions. The study leader must develop project-specific questions informed by the risks and other information gathered during the preparation stage.

The ensuing debate generally leads to the identification of higher level, more strategic risks than had been voiced in the initial risk listing. It also raises the debate to a level which senior management

**Table 8.2** Questions to identify strategic risks

Relating to business issues
　Is funding in place?
　Are the conditions right for the project to proceed?
Relating to the brief
　Does everyone understand exactly what the client expects to build?
　Do you have a current written brief and is it clear?
Relating to roles and responsibilities
　Is the structure of the client body clear?
　Are all appointments in place?
　Who is responsible for coordinating between the different disciplines/packages?
　Are all information/drawings consistent?
Relating to communications and decision-making
　Is there a management/communication plan and project directory?
　Are decisions made in a timely manner?
　Do you understand and use the process for change control?

finds interesting. Having identified these high-level risks the study leader will then ask the team to identify action owners, decide what they are going to do to mitigate the risk and identify a date by which it will be done. It is good practice to name an individual as the action owner rather than an organisation since this improves accountability. A short workshop will only address management actions for the most severe or urgent risks. The study leader should agree a plan to identify management actions against the rest of the initial risk listing at a later session outside of the workshop.

*Stakeholder assessment*

Next the workshop will address the stakeholder map. The study leader will lead a debate to identify who will liaise with the stakeholders with a view to maximising support for, and minimising resistance to, the project.

*Review of issues and observations*

Before concluding the workshop, the study leader should review all the issues and observations that he has recorded during the workshop to ensure that they have all been addressed. Now is the time to identify who is going to do what to resolve any issues that still open.

*Wrap-up meeting*

At the end of the workshop the study leader will summarise the outcome and get the team to agree on the next steps. If extra time is needed to resolve some of the issues he may arrange a wrap-up meeting at a later date. The wrap-up meeting can be part of a later project team meeting. This minimises disruption to the project team's working schedule and normally attracts greater buy-in than 'yet another meeting'. It may be better to call a separate meeting. It provides an opportunity for senior management, who were not present during the workshop, to contribute to the decision-making panel.

*Report*

The study leader will then compile a report similar to that described in the stakeholder conferencing workshop above.

A long established, publicly quoted company needed to refresh its image to remain competitive. Part of its strategy for doing this was to relocate its headquarters from a prime central city location to an up and coming fringe location. Not only would this facilitate the introduction of new working practices but it would provide an opportunity to generate significant value from the sale and redevelopment of the prime city centre site. The strategy to maximise value from the city centre site was to sell the long lease on the site to a developer after receiving planning consent. This strategy meant that they would need to evolve a brief for their project design team that could not be bettered by a prospective developer. The company commissioned a project definition study to assess ways in which the team could maximise value and minimise risk while, at the same time, evolve a plan for gaining the support of the influential external stakeholders in the project, notably the principal landowner in the area, who was also their landlord, and the local retail and resident associations. The study involved a few weeks' consultations to provide input information to a workshop attended by senior managers and their professional advisors. The outputs from the workshop were an agreed value driver model to form the basis of the design brief, an analysis of the main risks with a management action plan and a stakeholder management strategy.

*Outputs*

Project definition studies will have the following outputs:

■ Confirmation that the project is likely to fulfil the needs of the business.

- A functional model of the project linking the project aims and objectives through the value drivers to the issues that need to be considered by the design team (a value profile), thus informing the project (or design) brief.

- A high-level risk register and a formal risk management process.

- Identification of key stakeholders and the basis upon which their expectations will be managed.

- A basis for decision-making and the means to measure improvements in value as the project proceeds.

- A forum in which all participants gain a greater understanding of the project, its aims and objectives, and the issues and concerns of other members within the project team. This leads to greater collaborative working and an improved project outcome.

## Brief development

As the design develops, the team may evolve several options, each of which satisfy the requirements of the concept brief. They need to explore the feasibility of each of these options and to determine which is the best fit with the client's requirements. Selection between two or more good options can be complex and a value management study can help to inform that decision. The output will form the detailed design brief.

### Preparation

The brief development study will begin, like all studies, with a briefing meeting between key stakeholders to confirm:

- That the project objectives are unchanged and that the value drivers are the same as those developed earlier.

- A common understanding of the options on the table and precise objectives of the study.

- The information that is available, what might be needed and whether there is any reconciliation required between the different documents (e.g. cost estimates may not necessarily refer to the latest set of drawings).

- The procurement options under consideration.

The study team leader will be responsible for gathering this information, analysing it and distributing it to the participants before the workshop.

## Workshop

The workshop will begin in the usual way with introductions and confirmation of the objectives. There will be a workshop team briefing at which key players table any new information and explain what they are seeking the workshop to address. The next stage will be to review the value profile to confirm that project objectives and the value drivers are unchanged and that their weightings remain the same. Any changes will indicate a changed project which must be communicated to the project team.

## Function cost analysis

The study team leader should have gathered or generated whole-life cost models for each of the options under consideration before the workshop. The whole-life costs should be distributed across the function analysis before or during the workshop.

## Value drivers as evaluation criteria

Since the value drivers are those things which must be delivered in full if the project is to be a success, it is logical to assess each of the options against them. Other criteria for assessing the options will be the key project risks identified in the previous stage.

An effective way to assess the options with the team is to use a weighted option selection matrix (see Section 10.10). Using the weightings against each of the value driver as previously agreed, the study leader will work with the team to obtain consensus on the preferred option(s).

## Value for money

Any value driver directly relating to the whole-life cost should not be used as an assessment criterion. The reason for this is that having established how well each of the option satisfies the requirements of the other value drivers, dividing the resulting value index by the net present cost gives a measure of value for money. One of the options may be far and away the best in terms of satisfying the value drivers. It may, however, be extremely expensive and unaffordable. It may not represent good value for money.

## Sensitivity

It is essential to undertake sensitivity analysis on the results to test the robustness of the outcome since the techniques described above rely, to some extent, on subjective judgements.

This does not undermine the benefits of doing it with the whole team present, since it stimulates discussion. Everyone present learns significantly more about each of the options, its merits and its demerits. It also provides an audit trail as to why the preferred option was selected. Sometimes, the method will demonstrate that one option is, in fact, wholly unacceptable.

An example of this occurred during a programme of studies to establish the best options for an airline to occupy space at an airport. The airport offered space at several terminals each with different characteristics. One of the key value drivers identified by function analysis was to provide capacity for future expansion. An option appraisal had identified one of the terminals as having a significant advantage from an operational standpoint. Consideration against the value driver of 'provide capacity for future expansion', however, quickly ruled it out as a practical proposition because it offered no possibilities for expansion.

## Option improvement

Having selected one or more preferred options the workshop will then generate ideas to improve the selected option. This provides an opportunity to incorporate certain attributes of a rejected option into the preferred option. The techniques used are described in the section titled 'Value Engineering'.

## Report

The report will summarise the workshop proceedings, detail the selected option and the agreed proposals to improve it. Appendices will contain the workshop input documentation, any analysis of it and full details of all ideas raised for improving the selected option. The report will inform the design brief and the gateway 2 review on public sector projects.

## Procurement route selection

A method described above can be used to decide the preferred procurement route. In this case the assessment criteria will be weighted towards the key project risks and the preferred allocation between client and supply chain.

During the feasibility stage of the upgrading of a foul water treatment works in southern England, the consultants had narrowed the options of secondary treatment methods down to three BAFF, ASCAT and conventional percolators. All methods were proven and offered very little between them in terms of whole costs. The conventional percolators were ruled out due to space constraints. To select the preferred process, the value engineering team used a standard decision-making/evaluation matrix. The workshop team established the evaluation criteria as:

■ Sludge quantity and quality

■ Business rates

■ Reliability

■ Robust process

■ Ease of construction

■ User friendliness

■ M & E costs

■ Construction cost

■ Operating costs

■ Energy cost.

First of all, each member of the team ranked the criteria, excluding cost criteria on a percentage basis; the total for the team was then summed to form an overall ranking of the criteria. The two favoured options, BAFF and ASCAT were then rated against each criterion on the basis of:

4 = Very good

3 = Good

2 = Satisfactory

1 = Poor

For each option, the rankings were then multiplied by the ratings and summed; this gave a comparative quantification of the 'satisfaction of needs' or 'utility'. This latter figure was divided by the total cost of each option to give a final value index. The outcome was that BAFF was strongly favoured as the process to be adopted. A sensitivity check was then undertaken. This confirmed BAFF as the favoured option.

## *Value engineering*

Once designs have been developed to a point at which a reasonable cost estimate can be related to the design (e.g. Stage D on the RIBA scale), the team should undertake a value engineering study. The main purpose of the value engineering study is to optimise costs, delivery time and the quality of the proposed design.

The stages named below relate to the key process stages introduced in Section 2.6.

### Preparation (Stage 1)

A value engineering study will begin with a briefing meeting with the key players. These will now include a larger proportion of technical people than the earlier studies. During the briefing meeting the first thing will be to confirm that the project still has the same objectives and that the value drivers are relevant. The next will be to understand what problems, if any, the study needs to address. Sometimes the perceived problem is merely a symptom of a different underlying fault. During the briefing meeting the study leader should ask three questions :

1.  What is the problem (or opportunity) this study needs to address?

2.  Why do you consider that this is a problem or an opportunity?

3.  Why do you believe it is necessary to find a solution or, what are the consequences of not resolving this problem?

The study leader should explore the broad areas of opportunity for improving the project, the team's concerns and the constraints within which they are working. He should establish the scope of the study and what elements of the overall project may be excluded.

The study leader should establish the validity of constraints and confirm that they are not self-imposed constraints due to corporate inertia designed to stifle discussion of an area of the project that they would prefer not to discuss.

Workshop inputs will typically comprise:

■   A summary of the project brief

■   The current design report

■   Key drawings

■ Specifications

■ The cost estimate (preferably presented in a way that can be distributed to the function model)

■ The function (cost) model developed earlier.

## The workshop (Stage 2)

It is likely that by this stage in the project most of the participants will know each other well; therefore, introductions are likely to be largely unnecessary. However, it is still important at the outset to confirm people's expectations.

*Workshop briefing*

The workshop team briefing should be confined to new information and presentation of concerns, areas of opportunity and potential constraints.

The study leader should note any issues and observations raised during the briefing as described earlier.

## Function analysis (Stage 2)

The next step is to confirm the functional model developed earlier and the weightings given to the value drivers or primary functions. This confirms that the project has not changed. The team will then develop a more detailed function model down to elemental level, applying costs from the cost estimate. Both of these tasks should be done during the workshop to obtain the benefits of team learning through participation.

It is common practice in the construction industry to omit this step and go straight into generating ideas without reviewing or conducting a function analysis. This is a missed opportunity since it avoids generating ideas based on what things must do and thus stifles innovation (see Section 2.8).

## Risk review (Stage 2)

The team can now review the main risks to the project. This counterbalances the function analysis by making the team aware of what might harm the project. Identification of activities to reduce risk can be addressed in the next stage of the workshop.

## Creativity or generating ideas (Stage 3)

Using the function cost model the team leader will now invite the workshop team to generate ideas to address the workshop objectives. It is extremely valuable to ask the team to mark any functions where the cost seems high in proportion to their importance to the project before beginning to generate ideas. This can be done empirically or more methodically using the SMART approach described in Chapter 10. Ideas should be relevant to all the workshop objectives not just those to reduce costs. The most common way to generate ideas is in a brainstorming session (see Chapter 10 for this and other creative techniques). Using the function cost model as a 'route map' ensures that all areas of the project are explored for ideas to generate more value. If time is constrained the study leader can restrict the idea generation to those areas of the project offering most potential for improvement (as highlighted on the function cost model). A good brainstorming session may generate several hundred ideas.

## Evaluation (Stage 4)

Once all the ideas have been generated it is necessary to select those that warrant development into more detailed value engineering proposals. Evaluation techniques are discussed in Chapter 10.

Use of appropriate selection criteria ensures that no idea that has merit is totally discarded and can provide useful material for improved performance on later projects. The team select the best ideas and allocate owners to develop them into proposals.

It is acceptable to close the workshop at this point and agree on a date for a wrap-up meeting at which the proposal owners will present their recommendations to a decision-making panel a week or so after the workshop. The advantage of allowing the team to develop the proposals outside the workshop is that it allows them to gain access to their working files and to consult with suppliers and others who might be needed to make a contribution to the proposal. The worked up proposal is then likely to be much more robust than if it had been worked up during the heat of a workshop. The disadvantage is that the momentum of the workshop may be lost and proposals may not be developed.

## Proposal development (Stage 5)

The owner of each proposal is responsible for developing it within the agreed timescale and in the agreed manner (see typical proposal formats in Chapter 12). He will involve others in the team as appropriate.

## Presentation (Stage 6)

The owners should present their recommendations to the decision-makers in an agreed format (see Chapter 12). The recommendations should include a description of the existing design solution, a description of the proposed alternative, the advantages and disadvantages of the proposed alternatives and the impacts of the alternatives in terms of cost, time and quality. The owner may also express their views on how the proposal could be implemented and whether, having thoroughly evaluated the proposal, they still recommend it for implementation.

Value engineering proposals should be compared with a whole-life cost basis, since a small saving in capital cost could have a much larger impact on the whole-life cost of the facility.

*Interdependency and scenarios*

When considering proposals for implementation it is necessary to check for interdependency. For example, two proposals could be mutually exclusive. Failure to address interdependency can result in double counting of cost savings or other benefits.

An extension of this principle is to develop scenarios combining a number of selected value engineering proposals. A technique for constructing a scenario is given in Chapter 10. Essentially, each scenario consists of a grouping of selected proposals that might make up an improvement work package. The impact of each of these proposals on the achievement of each of the value drivers is assessed and the total impact on the project value index calculated. The scenario which has the greatest positive impact on the value index is the one that the team will recommend for implementation.

## Report (Stage 6)

The report will include a description of the workshop, a list of selected proposals and their estimated impacts on the project and appendices giving all the details of the information which went into the workshop, the analysis thereof and all the ideas and discussions which resulted in the workshop recommendations.

The output informs the development of the design, provides an audit trail for why decisions were made and is an input to gateway review no. 3 on public projects.

A value engineering study into a conventionally procured, proposed landmark office block in the city of London took place just before the commencement of detailed design. The client was concerned that the estimated costs were 10% too high.

During the workshop briefing it became apparent that the client's main concern was to improve the viability of his investment. This opened the way to reviewing ways in which to increase rental income as well as reduce costs.

The team developed a function cost model which identified the key value drivers as:

o   Create an exciting image externally and on entrance

o   Maximise the value of useable space

o   Optimise tenant occupation costs

o   Create a comfortable and secure environment

o   Accommodate third-party requirements.

A brainstorming session generated over 150 ideas of which about 50 were selected for development into proposals. The team agreed to work up the proposals outside the workshop and recon-vene for a wrap-up meeting about 1 week later. At the presentation, the team recommended 35 proposals for acceptance. The client agreed to implement most of them, resulting in cost reduction of 6% and an increase in NPV (due to increased rental) of a further 4%.

## Implementation (Stage 7)

The final stage in any study is that of implementing its recommenda-tions. The implementation plan should be the ultimate deliverable from any study. Without it the benefits of the study risk being lost. If it is not followed through, they will be lost. Ownership of the plan should therefore be entrusted to someone with the authority to monitor its progress, report its effectiveness and put in place remedial action if it is falling short of expectations. The need for implementation plans is stressed throughout this book.

## *Design and cost review*

Despite all the good preparation and application of value and risk management, when the cost estimates are market tested or the job is put out to tender, outturn cost estimates or bids may exceed budget expectations. At this late stage in the project all members of the team should be thoroughly familiar with the functional require-ments of the project and contributions made by each of the elements within it. It is therefore appropriate to apply a simplified form of study with the primary objective of reducing costs while retaining functionality.

This technique should not be used as a shortcut to value engineering. It is, however, considerably better than a straightforward cost cutting exercise which tends to reduce quality and functionality.

## Preparation

The pre-workshop briefing meeting is likely to be short to understand the nature of the problem, what information is available and to identify the primary functional areas of the project for consideration. Functional areas describe the uses to which different parts of the facility will be put once it is built. For example, in a hotel, the functional areas could comprise:

- Reception

- Bedrooms

- Conference facilities

- Back of house facilities

- External envelope

- Plant

- External areas.

It is normal to constrain the number of functional areas under consideration to about five or six.

## Workshop

The workshop team briefing is likely to be extremely short and confined only to new information previously not available to the project team.

The study leader will then select one of the functional areas and ask a member of the workshop team to give a short introduction to it. This will include the concerns, constraints that apply and opportunities which might be achievable for that functional area.

The workshop will then brainstorm ideas for improvement (usually cost reduction) but will not evaluate them.

The study leader then selects the next functional area, and invites a member of the workshop team to outline their concerns, constraints and opportunities for that functional area. They will then brainstorm it to generate ideas for improvement.

This process is continued without evaluation of any of the ideas until all functional areas have been addressed. It is common practice to have a final brainstorming session covering the whole of the project and any other ideas that people might wish to contribute.

Not until idea generation is complete will the study leader ask the team to evaluate each of the ideas to select the most suitable for further development into proposals. The method will be similar to that described above for value engineering.

The owners will develop the proposals in the same way as value engineering proposals and present these at a wrap-up meeting to obtain decisions.

The timetable for such an exercise is usually extremely short and the wrap-up meeting is likely to take place within two or three days after the workshop.

A short report will describe the workshop, the ideas that were selected, the recommended proposals and the decisions that were made at the wrap-up meeting for implementation. The report informs completion of the design, adjustments to the cost estimate and an audit trail for why the decisions have been taken.

> Market testing of the designs for a high-bay warehouse, to be used for storing chemicals indicated a likely cost overrun of 4.5%, shortly before tendering. A one day design and cost review identified opportunities to reduce costs by 6%.

## 8.3 Risk management studies

### *Strategic risk study*

At the inception stage, risk management efforts will be directed towards informing the business case in order to gain approval for a project to proceed. The study should be approached in the full realisation that the sponsoring organisation may prefer to derive the benefits it is seeking from a non-building project. For this reason, the risks to be studied will have a strong business, rather than project focus.

### Preparation

Adequate preparation is vital. It is likely to comprise consultation with senior management and other stakeholders to gather information on the main areas of potential risk. The following categories are suggested in the management of risk (M-o-R) guidelines in the Office for Government Commerce (OGC):

■   Strategic and commercial risk, for example, financial, demand, residual value, tax.

■   Planning and legislative risk, for example, site acquisition, consents, environmental.

■   Operating risk, for example, technological, obsolescence, continuity, staffing, availability.

■   Political/regulatory/legislative risk, for example, changes in law/government

■   Project specific risk, for example, access, design, construction

■   Force majeure.

The categories used may be stipulated by the sponsoring client to fit their internal procedures for assessing the business case. The study leader should select those that are appropriate for the circumstances in consultation with the client.

Preparation will involve consulting with the identified stakeholders, both internal and external, to gain their views and assessments of the most severe risks to achieving project success. Suitable forms for gathering and assessing risks are given in Chapter 9.

## Workshop

Unless it forms part of a wider stakeholder conference, a workshop at this project stage is likely to be short and involve only a small number of senior management. It may take the form of a well-chaired meeting as distinct from the stereotype workshops described earlier. On a large project it may be necessary to convene a series of workshops, each addressing a different aspect and stakeholder group. The workshop(s) will provide a forum for presenting and discussing the findings of the preparatory work and for assessing the main risks. These are likely to include broad-brush estimates of the costs and time impacts of the identified risks. During this meeting, it would be normal to agree on the broad allocation of risk between the likely contracting parties. In the case of PFI/PPP (PFI: private financial initiative) (PPP: Public Private Partnership) procurement the allocation would be between the public and private sectors and form the basis for the contract. Table 8.3 shows an extract from a typical risk allocation matrix as used in PFI projects.

**Table 8.3** Extract from a risk allocation matrix

| No. | Risk Heading | Description | Allocation | | |
| | | | Pubic sector | Private sector | Shared |
| --- | --- | --- | --- | --- | --- |
| 3 | Availability and performance risks | | | | |
| 3.1 | Latent defects in new build | Latent defects to the structure of the building(s), which require repair, may become patent | | ✓ | |
| 3.2 | Change in specification initiated by procuring entity | There is a chance that, during the operating phase of the project, the procuring entity of the services will require changes to the specification | ✓ | | |
| 3.7 | Availability of facilities | There is a risk that some or all of the facility will not be available for the use to which it is intended. There may be costs involved in making the facility available | | ✓ | |
| 3.9 | Force majeure | In the event of force majeure additional costs will be incurred. Facilities may also be unavailable | | | ✓ |
| 3.10 | Termination due to force majeure | There is a risk that an event of force majeure will mean the parties are no longer able to perform the contract | | | ✓ |

The risk mitigation strategies and the criteria for escalation should also be agreed. Owned and dated management actions for the main risks should be agreed.

## Analysis

The raw outputs from the workshop(s) will be analysed by the study leader and his team after the workshop and incorporated into a risk register. The cost and likelihood assessments will form inputs to quantifying risks and estimating appropriate risk allowances for the client's business case. At this stage, the study leader and his team may apply optimism bias to supplement the workshop findings. Details of this technique are described in Section 11.6.

## Report

The study leader will prepare a report, summarising the outcome from the workshop(s), the analysis of the results, recommended risk allowances and a risk register. The risk register will include details of the agreed management actions to address several risks. The report will also include agreed recommendations for the ongoing management of risk. It should be tailored to fit the client's business planning procedures.

A large PFI healthcare project involved a large number of stakeholders each of which needed to safeguard its own reputation for delivery of first class services. The initial strategic risk study comprised an anchor workshop, at which the risk categories and broad allocations between public and private sectors were agreed. This was followed by six other workshops, each involving selected groups of stakeholders, to identify assess and establish management strategies for the risks which might affect each of the stakeholder groups as well as the overall project.

## *Initial project risk study*

In Chapter 4 we introduced the concept of combining risk and value studies in project definition studies. Frequently, the initial project risk study will take place in this context. For clarity we describe the study as a freestanding study below.

## Preparation

The first task facing the study leader will be to convene a strategic briefing meeting with the objective of understanding the project, the issues that the need to be addressed, the perceived concerns and constraints, the key stakeholders and, last but not least, the expected outcome of the risk study. The study leader will consult with the key stakeholders to learn their views on the risks that could impact the project. During these consultations (which are best undertaken face to face) he should explore not just the project specific risks but also the other categories of risk referred to in Section 6.3, namely project, consequential, loss of benefit and business/political. The study leader should sort the risks into the agreed categories, (see above) and include the qualitative assessment of each risk. Because the information will have been gathered from several sources he will need to filter out duplications and inconsistencies in qualitative assessments when compiling the draft risk register. He should then circulate this to all those who are invited to the workshop.

## Workshop

After the initial introductions and workshop team briefings the study leader should validate the draft risk register with the team to ensure that his interpretation of the information gathered before the study is acceptable to the team. At this stage, if it is intended to quantify the risks, he should ask the team to agree on the parameters for quantification. The study leader then has two choices. He can either interrogate the register, probing perceived areas of weakness, using specifically aimed questions as discussed in Section 10.2 or he can address the risks in descending order of severity, as identified by their qualitative ratings. The former approach helps to raise the levels of the risks that are discussed. This is because it treats the risks that were identified in the draft register as symptoms of an overarching risk that may not have been identified explicitly.

Either way, having agreed what actions should be taken to reduce risks, he should identify action owners for each risk, the actions they will undertake to mitigate the risk and the timetable for undertaking those actions. The actions should be selected using the risk management strategies proposed in chapter 11.4. They should be added to the risk register.

## Report

The report will summarise the main findings of the workshop and include the risk register, complete with identified risk management actions, the ongoing plan for managing risk and, if required, an estimate of the risk allowances, calculated using the methods described in Chapter 11.

## *Risk reviews*

Although the high-level risk management process will have been started at the project definition stage, there may have been little attempt to develop management actions against all the risks that will have been identified. It is necessary therefore to undertake a review of all the risks shortly after the project definition study.

In addition, because the status of risks will continuously change as the project progresses, it is essential that risks are kept under constant review. Regular reviews of the risks and the actions taken to reduce them, in order to assess the effectiveness of the management actions and identify new actions, are not an option – they are essential. If risks have been quantified to create a risk allowance, the risk review provides an opportunity to update this allowance and to reflect on the effectiveness of the risk mitigation strategies.

There are likely to be two types of reviews, formal reviews and informal reviews.

## Formal reviews

Formal reviews should be undertaken at regular intervals, between milestones, by a 'risk management' group. The risk management group will have been charged by management to implement the risk management process. The intervals between reviews will vary from project to project. Wide ranging reviews should be undertaken at longer intervals, say between one and three months. There should be a formal review at all significant milestones. The procedure for a formal risk management review is as follows.

*Preparation*

A risk management review comprises more than just a workshop. Before any workshop, the study leader should make adequate preparation by asking each of the risk management team to do four things.

1.  Identify any new risks and assess their likelihood and impact.

2.  Identify any risks which are now passed (whether or not they have occurred).

3.  Make enquiries of risk action owners to assess the status of their actions and reassess their likelihood and impact.

4.  Gather any other relevant information.

The study leader should update the risk register and compile a workshop handbook, including the updated risk register, with the information gathered from the risk management team, inviting them to a workshop.

*Workshop*

*Briefing.* After introductions and confirmation of the workshop objectives (as described for other study types in this chapter), the workshop team briefings are likely to be confined to new information not previously shared with the team and particular concerns or constraints which may have changed since the previous risk review.

*Validation of updated risk register.* Immediately after the workshop team briefing the study leader will validate the revised risk register with the team. If it has not already been done this is a good opportunity to agree on the quantification ranges applicable to each band of qualitative likelihood and impact. The ranges given in the generic guidelines may need adjustment for particular projects due to the risk profile of the project.

During the validation of the risk register it is good practice to ask workshop team members if there are any new or additional risks, not gathered beforehand, that they wish to add. These are assessed and added to the register. Validation also provides an opportunity to challenge any of the assessments that have been made and identify any passed risks (things that no longer pose any threat because the project has passed beyond the stage when they could have occurred) that may still be on the register.

*Warning.* A risk, the likelihood of which has shrunk to negligible proportions but whose impact would be severe or disastrous should *never* be deleted from the risk register. Even though the project team may believe that it has taken every possible step to avoid or minimise a risk, there is still a small chance that it could occur. If it does, its impact will still be disastrous. Such a risk should be reviewed by the risk management group from time to time just to reassure themselves that it no longer poses a significant threat to the project.

It is not practical to review more than about 50 risks during a risk review effectively. It is therefore necessary to select those that come under active review against two criteria:

o   The identified management action should have been undertaken by the date of the risk review.

o   The severity of the risk indicated by the impact and likelihood.

o   The risk is imminent.

If this filter results in more than 50 risks, more than one review should be planned.

A very effective way to do this, if the risk register is kept on an Excel spreadsheet, is to simply sort the risks by decreasing severity and by date. The risk register may then be projected onto the screen using a digital projector so that the team can see the amendments as they are being made. The study leader starts with the highest severity risks and works downwards until the team reaches a pre-agreed threshold, below which active management of risks is not considered cost-effective, or until the team runs out of time. At least the most severe risks will then have been addressed. The risk management group should review risks that have not been addressed during the workshop informally (see below).

*Scope of review.* The review of each risk should address the description of the risk (has it changed in any way, is its meaning clear?).

o   Has the appropriate management action been identified, has an action owner been ascribed and an end date been identified?

o   If so, has the management action been undertaken, and was it successful?

o   Is further action required, what should be the revised assessment in terms of likelihood and impact?

When reviewing the risks additional risks may be identified. These may or may not be related to the risk under discussion. New risks should be added to the register and assessed in the manner described earlier.

At the end of the review, the recorder will print out the revised register and distribute it to all members of the team. This eliminates any delay or disruption to the project team's day-to-day activities.

*Report*

Following the workshop, if risks have been quantified, the study leader will work with the cost consultant and/or planner to re-quantify the risks according to the revised assessments and the agreed quantification ranges and arrive at a new risk allowance. This and the revised risk register will then be included with a short description of the review and a record of attendees to comprise the risk report. The report serves several functions.

■   It may be used to inform the contingency sums that need to be set against the project.

- It provides an assessment of likelihood that the project will be completed by the target completion date(s).

- It provides an audit trail for external reviews (such as gateway reviews).

- It may be used to inform the contract documentation to identify the allocation of risk between parties (this is particularly relevant for PFI projects).

- Identify risks that should be escalated to a higher management level.

## Informal reviews

An informal risk review should be undertaken at every project team meeting. Alternatively, the risk management group may review the risks on a regular basis and submit reports for consideration at such meetings. Each informal review would normally only address the top 10 or so risks, selected using the criteria described in the previous section. The risk management group should review the lower rated risks from time to time, to ensure that their status has not changed, as well as those on which management actions should be complete by the review date. The format of the risk review is broadly similar to that described above for the formal risk review. The report will simply be a section in the project report.

## Operational reviews – business continuity

Once a facility is in operation it is common to undertake a review of the risks that could jeopardise the operations that are undertaken within the building. This accords with the recommendations of the Turnbull Report. The focus of this type of review will be that of maintaining continuity of business operations. The review may be combined with a project review as discussed later.

## 8.4  Combining value and risk reviews

In the above sections we have in the main, described the value and risk management studies as though they were independent and separate events. In practice it is common, and the author would say essential, that a risk review accompanies every value management workshop. Likewise, every risk review should address and remind the team of the key value drivers behind the project. There is little point in spending time maximising value if risks are left unmanaged which could then undermine all the added value gained. By the same token there is little point in minimising risks in a project which fails to deliver the expected benefits. Failure to deliver the benefits represents the greatest risk to the client overall. In practice, the reviews are frequently run as separate events and some projects will exclude either risk or value reviews (and, worse, sometimes both!). These practices should be resisted.

## 8.5 Project reviews

From time to time during the currency of the project and certainly after the project is complete, it is wise to review performance with the project team. These reviews can enhance what is working well, address what is not working so well, and provide a useful check for measuring performance against key performance indicators (KPIs). A review of this type will also identify if any of the fundamental business drivers behind the project have changed. Reviews may take various forms. An extremely effective form is a facilitated event, involving preparation, a workshop and a report.

### *Preparation*

The study leader should hold a briefing meeting with key stakeholders to establish the objectives of the project review, the underlying reasons for it, and agree the format for gathering information.

For reasons we have given earlier, it is far more effective to gather input and information before any workshop than to try and amass it on the day. Typically, the study team leader will gather information by sending a questionnaire to all participants and asking responses to the following, or similar, questions:

- What has gone well, that you want to keep or enhance?

- What needs improvement?

- Do you have any suggestions as to how such things might be improved?

- What has not gone well and should be avoided in future?

- Complete any pre-agreed performance assessment against KPIs.

The study team leader will then either meet with, talk to or receive responses to the questionnaires and compile a résumé of the returns. He will submit this to the team along with an invitation to a workshop.

## *Workshop*

### Briefing

The workshop will begin with the usual introductions, confirmation of the objectives and short workshop team briefings outlining new information, concerns, constraints and observations. During these briefings, the study team leader will record openly the key points that are raised on flipcharts or some other medium to ensure that these are addressed later during the workshop.

### Successes

The team leader will then ask the workshop team to review the successes achieved to date (i.e. responses to the questions what went well and what they want to keep and enhance). By reviewing successes first, the workshop takes on a positive atmosphere. This minimises the likelihood of a blame culture emerging. For each of the positive returns the team is invited to explain the impact it has had on the project, why it was such a success and how it might be further improved. For each of the responses that require actions, the team will identify an action owner and a date by which any action should be completed or reviewed.

### Problems

Now when the team is feeling buoyed up with its successes it will be in a good frame of mind to address problems. Once again the study leader will lead the team through each of the responses, possibly grouped under appropriate categories, to identify first of all the impact, the underlying reasons for it being a problem or for the difficulties that have been encountered, and only then explore proposals for making improvements or bringing about change. For the significant problems, action owners will be identified and dates by which those actions should be completed.

The reason for exploring the impact and the underlying cause of each of the responses is that it helps the team to identify effective remedial actions. By assessing the impact of the problem on the project the team can judge the significance of it. By understanding the root cause of the problem (this may involve quite a bit of debate) the team can identify a more effective ways to improve or change the situation.

### Team buy-in

For a large group (as one might encounter in a partnering relationship involving members of the supply chain), it may be appropriate to subdivide the team into groups to address different aspects of the project review. The group work will be undertaken in exactly the same way as described above. A spokesperson from each of the groups will then present his group's findings back to a plenary session comprising all the groups. He would invite comment on the proposals and thereby gain the team's buy-in to whatever actions have been identified.

## Performance assessment

The final stage of the review would be to assess the team's performance against any KPIs that have been agreed. This assessment will include assessing the value index in the manner described under measuring value, Section 2.15. It would also include a review of the risk register and the current risk allowances.

## Report

### Outcome of the review

The report should record the activities undertaken during the review, list the outcomes in terms of the responses, impact, causes and particularly proposals to improve or change, action owners and dates by which those actions should be complete. It would list performance against the KPIs, preferably with a comparison to previous reviews, and it will include a report on the value added since the previous assessment and progress in the management of risk.

### Basis for continuous improvement

The report provides the basis for reporting group learning and performance improvement, particularly in projects based on partnering or framework arrangements where continuous improvement is one the underlying requirements of the team.

## Post-handover reviews

Once the project has been handed over for use, the team should undertake a project review. The review will assess the achieved realisation of benefits and the effectiveness of the value and risk management programmes in maximising these. Feedback from this review to the project team will improve performance on future projects. This is particularly important where project teams remain together for a series of projects. While the facility is in use there will be opportunities to improve the performance of the facility using value management and to minimise risks of disruption to operations by undertaking risk management and developing a business continuity plan. The format for these reviews will be similar to that described above.

## 8.6 European practice

In the Chapter 7 we have described a number of different formal study types that may be appropriate at different stages in a construction project's life. In some countries, for example, France, they favour a series of much shorter informal workshops addressing small chunks of the overall project, undertaken on a much more frequent basis than that indicated above. Such an approach requires great discipline, if the rigour of the value and risk management process is not to be lost. The risk is that the workshops become seen as simply an extension of normal design and project team meetings and lose their effectiveness.

## 8.7 The project cycle

The end of one project can trigger the beginning of another. Once benefits are flowing from a built facility there are two ways in which this might come about.

### *Process improvement*

The first is there may be a desire to improve the flow of benefits, or productivity by streamlining the business processes taking place within the building. Value management is a very effective tool for bringing about process improvement. So-called business process re-engineering is based upon the application of value management principles on business processes. The reader is referred to Chapter 7.5 describing the use of value management in improving the delivery of public services.

### *Facility upgrade*

The second is that, after a period of time, working practices will evolve, new technologies become available and markets will change to the point where the occupier identifies that what used to be a successful new facility, now requires upgrading or changing. This

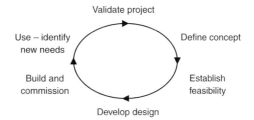

**Figure 8.2** The project cycle

can signal a start of a new construction project to further enhance the value to the sponsoring business.

This cycle takes several years to complete but with the increasing pace of change in society these cycles are becoming shorter and shorter. The information and communications technology project cycle is virtually continuous (see Figure 8.2).

## 8.8 Building on experience

Experienced teams apply the experience that they have gained over previous projects to the current situations, continuously improving the outcome. Building on experience in this way is what clients expect and deserve from their project teams. It should not, however, be used as an excuse for omitting formal studies, since it can risk incorporating outdated practices, missing opportunities for step changes in performance or failing to recognise significant risks. Structured programmes are always likely to enhance the quality of the service given by the project teams. This is particularly where they are working together regularly in framework arrangements delivering similar building types under which they are committed to continuous improvement targets. Ideally teams should keep an information base of the most effective proposals and actions arising from their formal studies to use on later projects. This will help them to meet their improvement targets.

### *The Kano principle*

A powerful principle at work here is the Kano quality model (Figure 8.3) put forward by Dr Noriaki Kano of the Tokyo Rika University in the late 1970s. Dr Kano integrated quality in two dimensions, the degree to which a product or service performs and

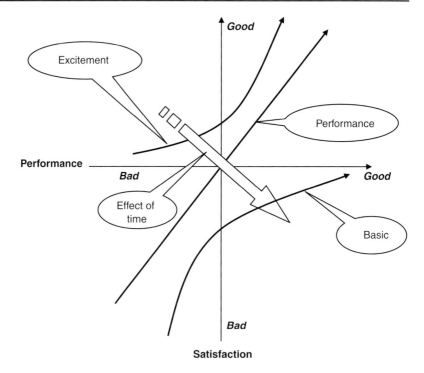

**Figure 8.3** The Kano model

the degree to which the user is satisfied. He identified three product attributes which one can assess against these dimensions.

1. Excitement quality
2. Performance quality
3. Basic quality.

## Basic quality

If a product does not contain certain basic requirements, it will not be suitable for its purpose. The customer will not be satisfied. In an office building, for example, occupants require certain minimum standards of lighting, heating and space to be able to work in comfort.

## Performance quality

Performance quality differentiators are, as the name implies, those things which may be present in greater or lesser measure and which might differentiate between two particular buildings, one of which has more or less of the performance differentiator than another.

An example of a performance differentiator here could be the space allocation for each office worker where 0.75 sqm/person is very tight (but workable) whereas 1.5 sqm/person might be generous.

### Excitement quality

Excitement quality features are those attributes which delight the user (delighters). These can include features that they never expected to find. In an office environment these could be a spectacular feature which causes the user to say 'wow'! when he enters the building.

## Migration with time

Over time there is a migration from the delighters towards the basics. Thus the delighters of yesterday become the performance differentiators of today and the basic requirements of tomorrow.

An example here might be air conditioning. Many years ago air conditioning would have been rare and stimulated feelings of delight among the users. As air conditioning systems became more established, some systems would work better than others. These provided a performance differentiator between different air conditioning types. In many office markets now, however, air conditioning is a must and without it the building would be 'unlettable'. Air conditioning has, in this context, become a basic quality requirement.

The migration from delighters to basic requirements can apply to any of the generic value drivers described in Section 10.8.

Formal value management studies can result in the creation of delighters through innovations in design. On the next project the delighter may be incorporated as part of the professional advice advanced to the client by the design team. Soon everyone is incorporating the feature and, without it, the building will be deemed substandard.

## Capturing experience

The Kano model provides powerful mechanism for teams to capture ideas emerging from the value management process and incorporate them within their value-adding services to clients.

# **9** Techniques for value and risk

## Summary

■ This chapter describes some of the more commonly used techniques that may be applied in either value or risk studies or both. They are also useful in many other day to day applications.

■ The technique of adequate preparation for any study is described in some depth because it is seen as essential for a successful study.

■ Workshops are a core element in most studies and the techniques of setting these up and running them is discussed.

■ No study is complete without a report so the essential elements of reporting are described. This is followed by a section on progress reviews.

■ The importance of basing studies on whole-life costs (WLCs) has been emphasised throughout this book. An introduction to whole-life costing techniques is given here.

■ Another common need is to engage stakeholders effectively. An outline of stakeholder analysis and management is described.

■ Another theme in this book is the need for people to work collaboratively together towards common aims. The chapter includes an introduction to partnering for this reason.

■ The chapter closes with a short description of mind mapping, since this is a useful way to capture and play back the sort of information used in many studies.

## 9.1 The need for techniques

No matter how attractive the concepts of value and risk management may appear, unless the study leader has a comprehensive range of effective techniques upon which to draw he will be unable to deliver results. This chapter describes some of the more commonly used techniques that the study leader might use.

Some of these techniques are common to both value and risk management, others have a larger role in one or the other. Many of the techniques are useful in other branches of project management. Many of them are workshop based (or may be applied during a structured meeting) others are to do with preparing the ground for a study or analysing information generated and reporting on it thereafter.

## 9.2 Preparation before a study

### Need for preparation

Both value and risk management utilise a workshop as a central part of most studies. However, there is a common perception that the workshop *is* the study. Frequently, workshops are held with little or no preparation. For example, where the practice of value and risk management is stipulated as mandatory, for example, as part of QA procedures, the project team will undertake a short workshop on each subject so that they can place a tick in the box to say that they have done it. Such exercises may be better than no risk or value management activity but in the long term they probably do more harm than good. Lack of rigour in the process generally means that it is less effective than it could be and, as a result, can give both value and risk management a bad name. The project team perceives the exercises as a waste of time and will therefore resist participating in value and risk studies in the future. A study should comprise a period of preparation, one or more workshops followed by reporting and implementation stages.

Like any effective and proven process, if either value or risk management is worth doing, it is worth doing properly. This means the application of a rigorous process in which the whole project team and other stakeholders contribute willingly and constructively to the study.

### Briefing

The first stage of any rigorous exercise is to ensure that all involved are properly briefed and that adequate preparation has been made to gather and analyse the necessary information before-hand. Making this information available as the input to a workshop will make the workshop itself far more productive.

Workshops are an expensive use of resource, they generally involve highly paid and senior members of the management team and there-fore need to be run efficiently. Furthermore, senior management have very busy lives and do not want to devote long periods of time during a workshop undertaking work that could have been done by individuals beforehand.

## *The strategic briefing meeting*

Before embarking on any study, the study leader should convene and chair a meeting attended by the key stakeholders in the study. Participants will depend upon the stage in the project at which the study is taking place. These are likely to include the following people:

- The individual who is commissioning the study

- The client who is commissioning the project (or his represen-tative)

- The user(s) – the ultimate users of the building or their representatives

- The project sponsor and project manager (if appointed)

- The design team leader (if appointed)

- Any other key representatives, key advisors to the client such as the agent or legal advisor.

This initial meeting should be limited to about six people.

The purpose of the strategic briefing meeting is twofold:

First, to enable the study leader to understand the project and stakeholders' expectations from the study.
Second, to agree on the type of study to enable the appropriate information to be gathered and analysed.

### Understanding the project
*Project objectives*

If they have not already been set out in a briefing document prior to the strategic briefing meeting, the study leader should establish a precise statement of the project objectives, expressed in terms of the benefits that the client body is seeking from executing the project. In the early stages of a project, defining the project objectives clearly and unambiguously may not be easy. The study leader should agree on the wording of this definition during the strategic briefing meeting.

*Success criteria*

The study leader should establish what will make the project a success, the success criteria, how these will be measured, and what targets, if any, should be or have been set. These will lead directly to the value drivers.

*Scope, stage and timing*

He should also understand the precise scope of the project to be studied (this may cover only part of the whole scope of the project). He should understand what stage the project is at in its lifecycle, and essential data on the timing, key milestone dates and the reasons for them.

*Costs*

At this meeting the study leader will explore what cost information has been generated, the business case that has been prepared to justify the project (if any), the basis for whole-life costing (see Section 9.9) and other issues which are of importance to the client body.

*Value drivers*

This discussion will provide a basis for developing a draft value driver model (function analysis) which will underpin any value and risk management studies following this briefing meeting.

*The reason for the study*

The study leader should understand clearly why the study is necessary and what the principal concerns of the key stakeholders are. This discussion should address the constraints within which the project is undertaken and which may apply to the study.

## Scope of the study

*Why have the study?*

The study leader should now have a fair understanding of the project and the reason for undertaking a study. The first stage in understanding the study details has to be to understand the nature of the problem. This assumes that there is a problem. If we define a problem as 'the difference between where we are now and where we want to be' we can see that problems do not have to be negative, but are statements of opportunity. Value and risk management studies should be planned into the master programme from outset of any project. It may be, therefore, the study for which the team is preparing has been planned all along.

*Things that the study will not address ('givens')*

There will be some subjects which fall outside the scope of the study. These may be things which have been set or agreed as policy by the organisation commissioning the study. They may be legal or political constraints which cannot be changed. There may be issues that derive from a wider project which are not within the scope of this study. These constraints are often described as 'givens' and should be clearly expressed at the outset of any study. This will avoid wasted effort during the study.

*Concerns and constraints*

The study leader will want to explore whether the participants at the meeting have any concerns about the project generally and where they perceive there to be constraints on its ideal development. These will include any deadlines by which the outcome of the study is required, for example, the next client approval milestone.

*What if the study cannot resolve the problem?*

If the study has been convened to address a specific problem, in order to understand the seriousness of the situation and possibly to focus the minds of those present, the study leader may wish to explore what would happen if the study team were unable to achieve the outcome which is expected of them. If the consequences are serious, this will focus minds on achieving a successful outcome.

## Study objectives

Now is the time to establish the precise objectives of the study. The aims of value and risk studies may differ in detail but are complementary. For example, at concept stage, both value and risk studies will aim to inform the development of the project brief. At design stage they will both aim to make the design solutions more effective, both to meet the project objectives and to minimise uncertainty. It is therefore entirely consistent that, while we have described value and risk studies separately in Chapter 8, because they involve different processes, they can and should be run together as single events.

The detailed aims of the value part of the study are likely to be expressed in terms, such as

- Clarify the project brief.

- Select the best options to inform design development or the design brief.

- Improve the financial viability of the project (either by reducing costs or increasing revenue or both).

- Establish the effectiveness of implementation plans.

- Resolve identified problems (including the occurrence of risk).

By contrast, the detailed aim of the risk part of the study is likely to contain phrases, such as

- Establish or refine a risk register.

- Establish or assess the effectiveness of risk management actions.

- Estimate the risk allowance.

- Establish the likelihood of completing on time.

- Establish how risk will be actively managed during the project.

Cutting across both the value and risk strands of the study, there may also be study objectives relating to the way the team and others can work effectively together.

## Setting targets

Each of the objectives described above should be expressed in terms of the scope, the success criteria (how we will know if we have achieved the objective) and, for most targets. Targets should be SMART, that is to say:

- Stretching

- Measurable

- Achievable

- Realistic

- Time bound.

## Avoiding failure

Stretching, but achievable, targets are not always popular if the team perceive that they will have failed if they fail to hit the target. The key is in the word 'achievable'. There is no point in setting targets that are neither realistic nor achievable. To avoid the sense of failure, the team should compile a graded 'shopping list' of proposals ranging from those that have a beneficial impact on the project (low hanging fruit), through those that have minimal (and easily acceptable) adverse impact (low pain) to those which may have a significant adverse impact (high pain but acceptable if the alternative is to cancel the project).

## Unwarranted optimism

There will be some instances where targets must be achieved by the team if the project is to be viable. In these instances it is wise to set study targets which exceed the minimum requirements by a significant margin. This is because some proposals will not achieve the optimistic outcomes calculated during the study. There is nothing worse than to promise to deliver significant savings, have the budget cut, only to find that the expected savings are not achievable in practice. Optimism bias (see Section 11.6) is a way to reflect this tendency in risk allowances.

## Information

By now the study leader should have a very clear idea of what is expected of the study and what information will be needed to maximise its effectiveness. He should explore with the strategic briefing team what information is available, who has access to it, who is familiar with it and who is not, and whether it needs some analysis before presenting it as an input. The study leader will agree how and where the information may be gathered. This could be by interview, questionnaire or through a study of existing documentation. If certain pieces of information are not aligned, for example, drawings are not in step with the latest cost estimates, he should ensure that such information is reconciled before any workshop.

## Study agenda

Value and risk management are structured processes which are most effective if things are done in a particular order. Getting the agenda right is therefore important. It is the study leader's responsibility to establish the timetable for conducting the study and getting decisions made. This should cover the preparation stages before any workshop(s), the detailed agenda for the workshop(s) and any post-workshop activities. Any study should have a definitive conclusion at which senior management make decisions based upon the recommendations of the study team and the developed proposals.

## Implementation plans

The study should result in an implementation plan showing how proposals to improve value or reduce risk will be implemented, by whom and within what timescale. The plan should include review points (e.g. brief reports in the monthly project report) to assess the effectiveness of the activities and allowing changes to be made if necessary.

## Building decisions

On other issues, such as helping people to work together, there will be decisions that need to be made on how stakeholders will be managed. Where the study relates to procurement there needs to be a decision on which procurement route is favoured. Often some of these decisions cannot be made during a workshop. There is simply insufficient time. The decision-makers may not be present. There may have been insufficient time and resource to undertake a proper analysis of certain proposals to give the decision-makers the information they need in order to make a decision that will stick.

A common process to close out a study is to convene a wrap-up meeting a week or two after a workshop or a series of workshops. This gives the team sufficient time to develop their proposals more thoroughly. It leads to more robust recommendations. The wrap-up meeting is discussed later in this chapter.

## Reporting requirements

The strategic briefing meeting should address the nature of the report which the study leader will be responsible for issuing at the end of the study to provide a record of it. The detail of report writing is covered later in this chapter.

## Gathering and analysing the information

By far the best way to gather information is to talk to those people who have it, whether or not it is also captured in documentation. Unfortunately, this is not always possible and one may need to rely upon telephone conversations, responses to questionnaires or exchange of e-mails.

The most useful types of information to be gathered in preparation for a study include:

- Information about the project and its objectives enabling the study leader to build up a draft value driver model (high-level function analysis).

- Stakeholders perceptions of risks, consequences and their severity, enabling the study leader to populate an initial draft risk listing.

- Identification of learning points from previous work identifying what are successes and what has gone well and things that need improving. This enables the study leader to draft a pro-forma for reviewing and learning from the previous projects.

- Who is involved in the project, how to contact them, their roles and responsibilities and the management structure.

- Information on stakeholders attitudes and influence to enable the study leader to draft a stakeholder map.

- Issues, concerns and constraints that stakeholders want addressed during the study.

- Information on how the project will be run.

The study leader should identify all other relevant pieces of information with which participants should be familiar to ensure that they come to a workshop fully prepared.

## Selecting the team

Once the above information has been explored and understood, the study leader needs to discuss who should be involved in the study, whether as part of a workshop or part of the consultation process beforehand. Team selection is crucial to the success of the study, particularly if it involves a short intensive workshop where the interaction of those present will have a crucial impact on the dynamics of the workshop and the resulting outcomes.

## Selecting the venue

The venue for any workshop needs selection with care to ensure that the workshop will run smoothly without interruption. Many advocate that workshops should be undertaken on neutral territory, for example, in a hotel or conference venue. This is not essential provided the study team and the hosting organisation enforce strict discipline on the smooth running of the workshop and make available the appropriate facilities. The neutral, off-site venue does have the advantage of making the workshop a 'special event' which encourages greater interaction and more serious contributions. We discuss the rules of engagement and facilities for workshops later in this section.

## Preparation workshop

One approach to prepare for a study is to conduct a preparation workshop.

The main aims of the preparation workshop are:

- To define the problem that the study should address

- To identify the scope of the study and any 'givens'

- To agree and define the aims and objectives of the study

- To establish the key value drivers or attributes

- To prioritise and weight the value drivers or attributes

- To agree on the study contributors

- To provide any necessary training or pre-study briefing to the contributors

- To identify, gather and analyse the appropriate information

- To establish the study timetable and logistics.

These are essential inputs to any value and/or risk study or a study that combines both processes.

The methods used are the same as those described in the section titled, 'The strategic briefing meeting' but in a workshop format with all key contributors present.

## Pre-study training

On a major study involving a number of workshops, or a programme of events, it is advantageous that some or all of the participants receive training to enable them to contribute more effectively. The nature of the training can vary from awareness training to introduce contributors to the concepts and methods that will be used, to full-scale competence training to build up a group of study leaders within the project team. The study leader should build into the study programme time for assessing training needs and for delivering or arranging to be delivered any training thus identified.

## 9.3 Invitation to participate

Having completed the preparation as outlined above, which can take a period of several weeks, the study leader should issue a formal invitation to all participants comprising the following information:

- The title of the project.
- The location of the venue and timing of the workshops or other events.
- The purpose of the study.
- The scope of the study including any things that will not be discussed during the workshop (givens).
- An outline of the process and the approach.
- The agenda for the study (including any pre- and post-events such as meals or breakout activities).
- A summary of the information upon which the study will be based.
- A note of any preparatory work that participants should undertake before attending the workshop.
- A point of contact for raising any queries.

### *Workshop handbook*

Depending upon the culture of the organisation and the familiarity of the study team with the processes that will be used, the study leader may wish to include outline details of the processes and how the team can best contribute to them. The study leader should agree with the client body whether this, more detailed, kind of workshop handbook will add value to the study. Swamping workshop participants with an abundance of paper before the workshop does not necessarily demonstrate thorough preparation. Regardless of the type of workshop invitation that is used, it should include copies or sources of any information that has been analysed and processed specifically for the workshop.

The workshop invitation should be distributed to the participants at least a week before any workshop to allow people to have sufficient time to digest the information within it. If this is the first indication to many of the participants that they are invited to the event, it should be sent out considerably in advance of one week to allow time for them to enter the event into their busy diaries.

## 9.4 The workshop

### *Workshop or meeting?*

Facilitated workshops are a very effective means of processing a large amount of information within a short time. The use of workshops have become very widespread but there is sometimes little to distinguish between a meeting and a workshop. The term workshop in this book refers to a facilitated event, following a period of preparation, information gathering and analysis, and following a predetermined, structured process, leading to a clear outcome or conclusion.

### *Place within a study*

There is a view that a workshop comprises the entirety of a value or risk study. The author does not share that view, particularly with the pressures to reduce the time allocated to workshops to the minimum. The workshop is the culmination of a period of thorough preparation and careful analysis at which the team process the input information to achieve a stated objective.

Workshops are generally intense events which are designed to encourage all participants to learn from each other and build on that learning to contribute constructive inputs leading to consensus decisions. The outputs are, therefore, true team efforts.

### *Duration*

There was a time where management were prepared to spend 4–5 consecutive days engaged in a structured workshop. This gave time for the team members not only to identify and agree what had to be done, but also to develop the proposals and actions in sufficient detail to enable decision-makers to take informed decisions on the way forward. Nowadays, workshops tend to be much shorter.

Long workshops have the advantage of utilising the full energies of the workshop team to work together in one place to build up their proposals. They also ensure that the participants come to clear conclusions and recommendations. A disadvantage, however, is that proposals may be developed in a rush, under non-ideal conditions (e.g. away from sources of information needed to fully process the proposal). Such proposals may later unravel when they are fully validated at a later date.

Workshops are likely to involve senior people. They are therefore expensive events, involving people who have many other calls on their time. As a result there is great pressure to use the workshop time as effectively as possible and to minimise its duration.

The scope of work to be covered in a study remains the same, regardless of the length of the workshop. It is clearly not possible to process large amounts of information effectively in a very short space of time. If a short workshop is to be effective therefore, the study leader must allow more time for thorough preparation beforehand and a clear plan for processing the information generated afterwards. This way the workshop itself can focus on just those parts of the study that require the whole team to be together. The workshop is retained purely for processing the information gathered and selecting the outputs for development later.

Regardless of the duration of the workshop, it is vital that it ends with a closedown stage. Otherwise the team will disband with many ideas still at large and numerous loose ends with no conclusions.

If it is not possible to plan the duration of a workshop to finalise decisions on its outcome, an interim closing down stage must be selected. For example, this could be after the evaluation stage in a value engineering workshop, when actions have been agreed, owners allocated to them and a timetable for those activities have been set. At least now the team will disband with a clear set of directives and time-bound tasks. Few decisions have been taken because the information has not been generated upon which to base such decisions. Under these conditions it is normal to call a 'wrap-up meeting' a week or two after the workshop at which the workshop team members can present the outcomes of their actions to a panel of decision-makers.

Regardless of the duration of the workshop, the study should cover all the stages of the process, without short cuts.

## Setting up a workshop

It is the study leader's responsibility to ensure that a workshop is set up in such a manner that it facilitates the processes that will be used and provides the ideal working environment.

*Timing.* This will have been determined during the strategic briefing meetings and should be clearly stated on the workshop invitation.

*Duration.* The study leader should resist pressure to reduce the workshop duration below that which is necessary to properly address the scope of work. Any workshop of less than half day duration is probably too short to be useful. Workshops of this duration should be confined to strategic workshops involving senior management. Workshops involving detail, such as value engineering workshops, are likely to involve at least 2–4 days effort. If some of the activities are undertaken outside the workshop (e.g. developing value engineering proposals) the workshop period may be reduced.

*Venue.* The study leader should agree with the client whether the workshop should be residential, held in a neutral location on the site or in one or other of the participant's offices. The space needs to be booked in good time to ensure its availability. It should be selected with a view to the adequacy of the facilities (see below) and the study budget.

*Room layouts.* In selecting the facility, the study leader should consider whether there are likely to be breakout sessions and to ensure that adequate space is allocated for these, either within the main room in which the workshop is held, or in separate rooms. The positioning of the furniture will have a marked effect on the workshop atmosphere. A classroom style layout is very formal and maybe suitable for training events where all eyes are upon the trainer, but is not conducive to interaction between the workshop contributors. Boardroom style layout is good for interaction between the contributors but is not so good for interaction with the facilitator, since he spends much of his time talking to the back of participants' heads! A U-shaped table encourages plenty of interaction between workshop participants as well as the study leader, who can move within the U and speak to people face to face. A very effective layout for informal workshops, with lots of interaction comprises several tables, of up to six seats each, distributed around a large room, each with its own flipchart or other visual aid, and supplemented by a central table for use by the facilitator and his team for processing and projecting plenary information.

*Visual aids.* Conveying information is crucial to the success of any workshop, so good visual aids are vital. Modern projection facilities direct from a laptop computer provide an excellent medium for projecting the information, not necessarily just PowerPoint slides but also processed information in the form of Excel worksheets and the like. However, they are relatively inflexible and certainly do not allow contributors to write their own notes upon the screen! Flipcharts are an excellent means of communicating information. These allow workshop team members to contribute their own additions to the information which is being processed. Such interactive contributions enliven the event. Overhead projectors are a very good means of conveying previously prepared information. OHP slides can be written upon during the workshop and provide good interaction. Whiteboards are not as flexible as flipcharts, because the information needs to be erased or copied onto an A4 sheet prior to erasure whereas flipcharts can be stuck up around the room. (The study leader should always check with the owner of the venue to ensure that sticking things to walls is permissible) Whatever visual aids are used the lead facilitator should be expert in their use and preparation (see the section on presenting information). It can be very frustrating for workshop participants if the facilitator is not fluent with the visual aids that he chooses to use.

*Workshop specific visual aids.* Large (A0 size) copies of pre-prepared (such as a draft function diagram) can provide a very visual and interactive way of conveying information, absorbing feedback, amending that information in full view of the contributors. This helps to build consensus agreement to a finalised and accepted version of the document. The advantage of a large printed sheet is that it can remain permanently on display throughout the workshop complete with the amendments and additions which may have been made.

Where there has been little opportunity to assess the likelihood and impact of risks before a workshop, a large colourer risk matrix (or heat map) is a very interactive way to assess risks quickly and accurately.

*Support facilities.* A workshop should be conducted in a room which is free from interruption. It is, however, necessary from time to time to communicate with the outside world or make photocopies. There should therefore be a telephone, fax and photocopier available for the use of the workshop participants. A wall clock is a useful aid to ensure that proceedings keep to time.

*Refreshments.* As Napoleon is reputed to have said 'an army marches on its stomach'. A workshop will run much more smoothly if participants are in a comfortable environment with adequate refreshments served. The study leader should ensure that the venue has the facilities to serve tea, coffee and other light refreshments as well as the ability to serve a buffet lunch. Serving a buffet lunch in a separate room can provide a welcome break. Continuing to work through a working lunch over papers in the workshop should be avoided unless absolutely necessary. A break at lunchtime and indeed at mid-morning and mid-afternoon, preferably accompanied by walk outside in fresh air (giving people an opportunity to use their mobile telephones) will help participants to maintain clear heads. Be useful. Arranging some social activities, such as dinner the night before a workshop, or breakfast on arrival enables people to meet each other in a friendly and relaxed atmosphere. This is particularly welcome if you have invited people from afar for an early start. Do not serve large meals during workshops as this will interfere with concentration after the meal.

*Accommodation*. Where a workshop is to be residential or if people have to come from far and wide, the study leader should arrange accommodation and suitable hotels either nearby or, ideally, on the same site.

*Entertainment*. Where a workshop is part of a broader team-building activity it is common practice to arrange some kind of entertainment usually in the evening but possibly during a break during the day. The study leader is responsible for arranging this, professionally, including preparation to ensure that the entertainment runs smoothly. It will be counter-productive if an event that was planned to encourage interaction between workshop participants has precisely the opposite effect.

## Active listening

During presentations by members of the team in a workshop, the study leader will record key observations and issues on a flipchart for all to see. This requires the skill of active listening – the ability to concentrate on what is being said, and interpret and summarise it in a few words. The skill is not significantly different from that needed when taking notes during a meeting except that he needs to be more concise and accurate and write the notes neatly on a flipchart. From time, or if he is not certain that he has understood correctly, he should summarise the notes he has made to the team and make any corrections they may request.

## 9.5 Recording

Workshops can generate huge amounts of information within a short space of time. This must be captured and recorded if full advantage is to be taken of the workshop environment. The study leader will be largely pre-occupied with ensuring that the process is followed by the participants rigorously and effectively. While he may make a written record of participants' contributions, these should be gathered at regular intervals and recorded, preferably on a laptop computer, by a competent recorder. In a large workshop comprising of several tens of contributors, more than one recorder may be needed. It is the study leader's responsibility to ensure that the recorder is thoroughly briefed and familiar with his duties which may be summarised as follows:

■ Attend briefing and wrap-up meetings as well as workshops

■ Be familiar with the workshop process

■ Prepare and send out invitations and make arrangements for workshops

■ Take notes and gather information that may be presented or handed out

■ Record entries on flipcharts or other visual aids

■ Compile reports.

When building up a resource of trained facilitators within an organisation, as described in Section 4.10, involving fledgling facilitators as recorders in workshops run by more experienced facilitators can provide excellent experience for them. The study leader should encourage the fledgling facilitator to co-facilitate some parts of the workshop to build confidence.

## 9.6 The wrap-up meeting

The wrap-up meeting should take place a week or two after the workshop, giving the team time in which to develop proposals. It will be chaired by the study leader and is used for two main purposes. First to obtain decisions on specific proposals which have been developed by the workshop team. The second is to formulate and agree on a clear plan for following through the value and risk management.

It normally lasts only a few hours and may be combined with a scheduled project team meeting. The decision-makers, key stakeholders and those who have developed proposals (the proposal owners) should attend. The owners will explain their findings and clarify any queries so that the decision-makers can make informed decisions. The study leader will record the decisions and the discussions surrounding them for inclusion in the report.

## 9.7 Reporting

The study report is an essential output from a value or risk study. Not only does this document provide an accurate record of the study for the team and third parties, but it provides an audit trail for future project reviews. The report will assist those conducting gateway reviews to satisfy themselves that the project team is ready to proceed to the next project stage.

There are two types or reports that are commonly used: progress reports and formal study reports.

### *Progress reports*

Progress should be entered into the regular project reporting mechanism such as the monthly progress report to inform management of the status of the value and risk implementation plans. Where benefits are not likely to increase by the amounts anticipated, or uncertainties are not reduced by the amounts anticipated, recommendations for remedial action should be highlighted in the project report. These would include proactive suggestions of ways in which the management team can help the owners of the proposals and actions overcome their difficulties. Such follow-up will thereby maximise the true potential of the value and risk management activities.

The typical structure of a progress report is indicated in Table 9.1.

**Table 9.1** Regular progress report headings

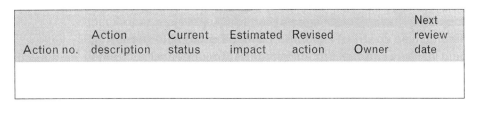

| Action no. | Action description | Current status | Estimated impact | Revised action | Owner | Next review date |
|---|---|---|---|---|---|---|
| | | | | | | |

## *Formal study reports*

In addition to the regular reporting structure outline above, each study should generate its own study report so as to inform management and the project team and act as an audit trail so that third parties can understand how and why certain decisions were made.

Readers of reports will generally be most interested in the outcomes of the study. They may also wish to satisfy themselves that the processes undertaken were robust. Formal reports may be aimed at different audiences. Some will be addressed to senior management to assist their decision-making processes. Others may be mainly for the benefit of the project delivery team. Some reports may be aimed at both parties or even a wider audience. The study leader should structure reports with the following key sections.

1.  A summary outlining the purpose of the study, the main outcomes and the implementation plan.

2.  A synopsis of the project, a short description of the processes followed during the study and who was involved.

3.  Appendices providing the detailed support to the outcomes, including the full value management plan and risk register, the information upon which the study was based and other relevant information.

The report may be bound in sections to allow distribution of the different parts to different parties.

## 9.8 Reviewing progress

From time to time in any project it is good practice for the team to get together to assess how the project is going, what lessons can be learnt and how these may be incorporated in a continuous improvement programme. These reviews could be triggered by several events.

At the end of one project, if the team is likely to be working together on another project, such a review may be a very good way to understand lessons learnt and improve performance on the follow-up project. The same review will inform the client how well the project has met his requirements and inform any post-occupancy improvement work that he may wish to implement.

A similar progress report format is useful at key milestones within a project to review past performance, learn lessons and come up with proposals to improve future performance.

## Briefing

The formal project review should be undertaken with the same rigour as the value and risk management study. The study leader should convene with the key stakeholders a strategic briefing meeting to understand the issues to be addressed, and their expected outputs from the study. At that same meeting the details of the timing and other logistics of the study will be agreed.

## Questionnaire

The study leader will then prepare a questionnaire to act as the basis for consulting with the key project stakeholders both internal to the project team and externally. He will then either request that they complete the questionnaire and send it back to him or better, interview them either face to face or by telephone to help them complete the questionnaire. The questionnaire addresses the following subjects:

- Successes and how to build on them

- Things that need improvement

- Things that were not successful and should be avoided in future

- Things that have not been used but maybe should have been.

Having gathered responses to the questionnaire the study leader should analyse the returns and consolidate the returns, grouped under the appropriate categories, to provide the workshop input documentation. A suitable format is shown in Table 9.2.

## Workshop

The study leader will then convene a workshop, usually of about one day duration and ask the team to work through the input documentation, category by category, filling in the root cause and impact before agreeing the action (if any) necessary to identify an owner to see it through.

**Table 9.2** Proforma for a project review workshop

| Feedback | Root cause | Impact | Action | Date by | Owner |
|----------|-----------|--------|--------|---------|-------|
| Completed feedback from project team member | Left blank for completion during the workshop | Left blank for completion during the workshop | Left blank for completion during the workshop | Left blank for completion during the workshop | Left blank for completion during the workshop |

Requiring the team to identify the root cause (of the feedback) and its impact will assist their understanding of the issue and help them propose an appropriate level of action.

It is usual to begin with the project successes and how to build on them, before looking at problems. This encourages a positive atmosphere and reduces any tendencies for recriminations. The study leader should prevent the team from going down the latter path so that constructive lessons are brought out. After the workshop the study leader will write a report summarising the team's recommendations.

## 9.9 Whole-life costing

The table in Section 12.14 outlines some commonly used headings and may be used as a checklist when compiling WLC estimates. The cost headings indicated in this table relate mainly to the costs of maintaining and operating the building.

By far the greatest contributor to the WLC of an asset is the business operating costs. These will be specific to the operation being undertaken within the building and are outside the scope of this book. It is, however, important to understand that the building will contribute to improved productivity and therefore have an effect on the business operating costs. Therefore, when comparing options and on a whole-life costing basis the team must understand what effect the different options may have on the business operating costs.

For anyone undertaking a building project, costs should be gathered under appropriate headings using current cost levels. The data is entered on a standard spreadsheet which performs a discounted cashflow calculation to arrive at the WLC, expressed as a net present cost (or value).

## Comparing future cashflows with capital costs

It is not possible to directly compare capital initial costs, periodic costs, annual costs and income costs. This is because the first two are one-off costs while the last two are represented by annual cashflows. The effects of inflation and the cost of money on future cashflows means that they will be worthless in the future than current price levels. To allow the value of future cashflows to be compared with current costs it is necessary to calculate their present day cost. This is done by using discounted cashflow analysis and is described later in this section. The same technique is used to calculate the present day value of all future revenue streams.

The net present value (NPV) of a project is calculated by subtracting the present day costs of predicted cost cashflows from the present day values of all estimated future revenue streams. A positive NPV indicates that the project will be profitable. A negative NPV indicates that it will not. The amount by which the NPV is positive gives a measure of how profitable the project will be (see calculation of internal rate of return (IRR) below).

## Discounted cashflow analysis (DCA)

This is a method of converting future cashflows to current monetary values. The method requires gathering information, at current rates, on initial costs, periodic costs and annual costs (see table of headings in Section 12.14). From this data it is a simple matter to calculate the predicted annual cashflows, expressed at current rates, but occurring in the future.

The annual cashflows indicate the total costs (and/or revenue) for each year throughout the life of the asset all expressed at current prices. In order to convert them to a present-day value (which directly compared with current capital costs) it is necessary to apply a discount factor. The discount rate depends on two pieces of information.

The first is the cost of money to the organisation that will be using the building. This may either be the cost of borrowing, or, in the case of a commercial organisation, the return they would expect on money that they invest. These figures can generally be obtained from the finance director of the organisation concerned. In the public sector HM Treasury stipulate a discount rate related to the social cost of capital, currently 3.5%.

The second figure is to agree an estimate of the rate of inflation over the life of the asset. From these two figures it is possible to calculate the real discount rate (a rate which effectively takes out the effect of inflation) as follows:

$$(1 + \text{real discount rate}) = \frac{(1 + \text{actual discount rate})}{(1 + \text{annual inflation rate})}$$

For example, if the cost of money for an organisation is 12% and the expected rate of inflation over the life of the asset is 3%, the real discount rate may be calculated using the formula:

$1 + \text{real discount rate} = 1.12/1.03 = 1.087,$

giving a real discount rate of 8.7%

To calculate the discount factor which should be applied to the cashflows in any year, the real discount rate is inserted into the following formula:

$$\text{Discount factor} = \left( \frac{1}{1 + \text{real discount rate}} \right) (\text{year no.})$$

where year no. = the number of years since the start of the development.

For example, suppose the cashflow in the tenth year of the asset life is made up as follows: maintenance and refurbishment costs £50 000, utility and energy costs £25 000, staff costs £100 000, other costs £25 000 = total cash outgoings of £200 000. The discount factor will be $(1/1.087)^{10} = 0.43$. The present value of the tenth year cashflow will therefore be £200 000 × 0.43 = £86 000.

The discount factor multiplied by the annual cashflow converts it from current price levels to a discounted present value. This is less than the current value and represents the amount of money one would need to set aside in order to generate the future cashflow at today's values.

## Net present value

Supposing in the above example, the asset generates £300 000 worth of revenue in the same year, the present value of the revenue will be £300 000 × 0.43 = £129 000. The NPV of the asset in that year is the difference between the present value of the revenue and the present value of the costs, or £129 000 − £86 000 = £43 000.

The sum of all NPVs from the annual cashflows gives you the NPV for the whole project.

If the project generates revenues in excess of the costs over the lifetime of the asset on a discounted cashflow basis the asset will have a positive NPV. Generally speaking, commercial organisations will not invest in projects that do not show a positive NPV unless the investment adds some other value, for example, conforming to new regulations or enhancement of the performance of some other part of the business.

Where the asset generates no revenues, the NPV will be negative. For ease of terminology this is sometimes referred to as the net present cost of the asset. Use of discounted cashflow is essential when comparing different options (as maybe generated during a value engineering study or in comparing different options at feasibility stage).

A restaurant operator was considering two proposals for redeveloping an existing site. The first entailed doing a minimum amount of work to give the restaurant a new image, refurbished the existing 'back of house' facilities and retained the existing capacity in terms of numbers of diners that it could accommodate. The second proposal involved demolishing the existing building and rebuilding it with a larger dining area and smaller, but more efficient 'back of house' area. Although the second proposal would cost more in initial building costs, it would cost less to run (because of the more efficient 'back of house') and could accommodate more diners, thus increasing revenue.

A whole-life costing exercise demonstrated that the second proposal had a significantly higher NPV, easily outweighing the higher initial cost. The effect, over time, was that the building works were free, paid for by the increased revenue and lower operating costs.

## Internal rate of return

The IRR is commonly used by businesses to assess the relative value of projects. It is calculated by increasing the discount rate used in the above calculations to the point where the NPV equals zero.

The initial cost of construction generally represents only a small proportion of the total NPV. There is an important lesson to be drawn from this, particularly if the prime focus of a value engineering exercise is to reduce capital costs. It would do the client no good whatsoever if one reduces capital costs only to increase certain areas of running costs. In the long term it is better to do the reverse to increase capital costs in order to reduce running costs. In practice, decisions of this nature are generally made at a very simple level against the payback period for the proposed increase in capital costs.

For example, if a 10% increase in the cost of air conditioning plant reduces energy consumption and results in a short payback of 2–5 years, many clients will opt for the increased capital expenditure.

> During a value engineering study on a hotel refurbishment and expansion scheme, the hotel operator had a fixed capital budget which was in danger of being exceeded. The value engineering team presented a proposal to spend an extra £500 000 on an energy recovery system. They explained that the payback period for the system was about two years due to the predicted energy savings. The operator accepted the proposal, paying for it from his operating budget.

When making comparisons between two alternative proposals (as in the above examples) small errors in the base data used for calculating the WLC tend to cancel out. If, however, the calculation is to be used to predict future profitability, the data must be accurate. This is the situation, for example, when using WLC to predict unitary payments under private finance initiative (PFI) or PPP schemes. Under estimation of unitary payments due to the use of wrong data could lead to catastrophic losses for the operating organisation, the private sector partner. In these circumstances it is necessary to refine the estimates of the base data successively as the scheme evolves and more information becomes available. This must be done before financial commitment is made (financial close). At the same time it is advisable to undertake quantitative risk analysis in order to provide robust risk allowances for future uncertainties.

## Sensitivity

The figures used in WLC calculations are estimates of the expected costs at some time in the future, they are uncertain. For example, the rate of inflation over the life of the asset is based on economic forecasts. These are notoriously inaccurate. The study leader should therefore run a calculation for a range of inflation values. Certain annual costs, such as the price of fuel, may be particularly uncertain in the future. The study leader should apply sensitivity analysis to the calculation for different assumptions on future fuel costs. Only if, after applying sensitivities, the case for one option is significantly better than another option should the WLC calculation be used as the main reason for making a decision between options.

## *Varying inflation rates*

A further sophistication along these lines is useful if it is expected that the cost of one commodity will inflate at a different rate from the general inflation figure that you have used in calculating the real discount rate. If, for example, the price of labour is expected to outstrip general inflation by say, 2% per year, you can take account of this in the model by simply inserting a second labour cost line in the spreadsheet which itself inflates at the rate of 2% per year. In a sophisticated model this technique can be applied to as many lines as is necessary.

## *Cost benefit analysis*

When assembling the business case for making a change, it is necessary to demonstrate that benefits outweigh the costs. The technique of cost benefit analysis involves identifying all the costs and all the benefits and then undertaking a discounted cashflow analysis in the manner described in the previous section. To do this it is necessary to convert the benefits to cash amounts. This is not straightforward with non-monetary benefits such as improvements in safety or image. Some organisations, such as the NHS or Highways Agency have developed tables of cash sums to set against commonly considered non-monetary benefits, such as saving a life. Where these are not available the study leader should arrive at a consensus figure with the team to use in his calculations. This could be benchmarked against agency norms such as those mentioned above.

A positive NPV indicates that the benefits outweigh the costs.

## 9.10 Stakeholder analysis

To engage stakeholders effectively, it is necessary to have a process to assess and position them. The method outlined below can be

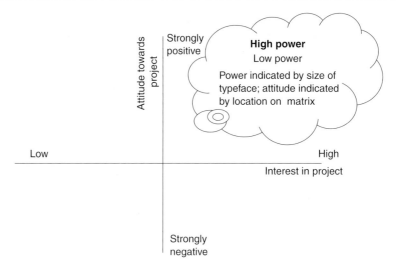

**Figure 9.1** A stakeholder matrix

used to understand the stakeholder environment and to prioritise resources.

The first step is to identify stakeholders. You cannot manage them if you do not know who they are. The study leader can identify stakeholders by talking with key project team members through the briefing stage of a value or risk management study.

Next it is necessary to understand the level of power and interest each stakeholder has to influence the project. This is not a precise art. It is based on perceptions. It should be based on their point of view. For example, a key fund-raiser may be of great interest to the project team, but is the project of interest to him? If the project does not proceed he can use his funds elsewhere.

Identify whether the individual stakeholder groups are broadly positive or negative about the project. Those with a high level of interest tend to have strong views. Those with little interest may not have strong views. Having established their interest and levels of influence, it is possible to plot the main stakeholders on a matrix as indicated in Figure 9.1.

The matrix informs the basic stakeholder management and communication strategy for the project.

## *Active management of stakeholders*

The aim of stakeholder management is to encourage the support of those who support the project and to minimise the influence of those who do not. Management effort is most effectively deployed to engage stakeholders with a great interest in the project and who have lots of power. Whether this group is for or against the project, it will be able to exert a lot of influence. It is therefore essential to make the effort to engage them actively. If they support the project the manager should:

■ Provide information to maintain their support

■ Consult with them prior to taking project decisions

■ Meet with them regularly

■ Consult with them, involve them and seek to build their confidence in the project and the team

■ Encourage them to act as advocates for the project

■ Nurture them and look after them. They are critically important to the project team and the project.

If they are against the project he should:

■ Attempt to develop their support and change their view by ensuring they fully understand the project and the benefits it will deliver. Their resistance maybe due to lack of information or understanding.

■ Attempt to build their confidence in the project and in the team.

■ Find out what is important to them. If he can help them out or minimise the negative impact on them they may be more helpful.

■ Demonstrate that he is doing his best to limit adverse effects on them.

■ Counter any negative influence they may have on others.

Another group worth spending time with are those with lots of power but with a relatively low interest in the project. These are the unexploded bombs – their interest is low at the moment. However, if the project alters or the individual's change, their interest may suddenly increase and they will use their power to influence the project.

With this group, the manager should:

■ Maintain a careful watching brief and make sure that changes to the project or changes within the stakeholder organisations do not suddenly increase the level of negative interest.

■ Find out what is important to these groups and make sure that the project does not adversely affect this. If it is likely to have a positive effect make sure they are aware.

■ Beware of other negative stakeholders trying to influence the stakeholder group to encourage them to oppose the project.

Stakeholders with a great interest in the project but little power can, if they are positive supporters, make strong allies. The manager should treat them well, provide them with information, involve them and use them to lobby other groups. If they are negative, they may deluge the team with e-mails and phone calls. The manager should not spend too much time on them. If they support the project he should:

■ Maintain their enthusiasm and interest in the project, they are good allies to have.

■ Provide them with information, invite them to presentations, involve them as much as resources allow. This can be done fairly cheaply through a project website, newsletter or open presentations.

■ Seek their input and opinion if practical since they will be flattered by this, but guard against getting too many opinions.

If they are against the project he should:

■ Respond politely but briefly to their communications, ideally by means of a standard letter or news bulletin.

■ The project sponsor or client representative may need to take a firm line with them as they can distract a lot of time and resource.

The final group comprises those with little power or interest in the project. They should not be given much management time. If they are supportive, provide them with information and be nice to them, their position or view may change in the future. Ensure they receive the project newsletter, have access to a project website or are invited to presentations.

# 9.11 Partnering

Partnering is relevant to value and risk management for two principle reasons. First, because the good working relationships inherent in partnering are also needed to deliver best value and minimise risk and, second, because running a partnering workshop requires similar facilitation skills to those needed by value and risk study leaders. Frequently, value and risk practitioners also provide partnering facilitation services.

The following section provides advice on facilitating a partnering workshop. These guidelines are written for the situation where the workshop comprises a single group up to about 18–20 in number representing the project stakeholders. It does not cover larger groups, such as those which are encountered in stakeholder conferencing, the techniques for which are described in Section 8.2.

## *Strategic briefing meeting*

The first step is to meet with the client and main stakeholders to establish:

■ The nature of the project

■ The key stakeholders and the politics of the situation

■ The individuals who should attend the workshop

■ The reasons for considering partnering

■ Issues and concerns expressed by all potential partners

■ Establish workshop details, such as the venue, date, agenda and reporting procedures.

The outcome of this meeting should be included in the invitations to the workshop.

## *Issuing the invitation*

Prepare an invitation to all participants, bidding them to a workshop. Agree with the client whether you or the client should distribute the invitation. This document should include the purpose of the workshop, the scope of the project, the agenda and notes from the strategic briefing meeting.

## Introductions

- Prepare place names to ensure that everyone knows who is who but let people sit where they want.

- Ask attendees to introduce themselves.

- Introduce yourself and your recorder.

- Make a *short* presentation of the approach to partnering that is to be adopted and this *specific* workshop's objectives.

## Presentations by key partners

Before the workshop advise any key participants who are expected to give presentations so that they can prepare the material. Encourage the use of visual aids/PowerPoint. Ask them to address the following issues:

- Their individual and their organisation's key objectives.

- Confine their subject matter to new information, the main issues to be resolved, their main concerns, gaps in information or perceived constraints.

- Avoid long descriptions of their organisation's or the project history.

The facilitator should allow discussion after each presentation for 5–10 min. Throughout the presentations and discussions he should actively listen and record issues and observations on a flipchart.

## Develop set of common objectives

Using the stated objectives from presentation as a starting point, facilitate a discussion/brainstorming session to identify common objectives. Write these on Post-it notes. With the team, group the Post-it notes under 5–8 main headings. Get the team to agree wording for each of the 5–8 common objectives (including qualifying remarks, targets, measures, etc.).

*Note:* The objectives should be measurable. Agree with the team on the Key Performance Indicators (KPIs) by which they will be measured.

**Table 9.3** Typical objectives, measures and targets for continuous improvement

| Attribute | Measure | Continuous improvement. Examples of targets |
|---|---|---|
| Corporate objectives | | |
| Make limited resources go further | Output per person | $+a\%$ p.a. |
| Improve certainty of outcome | Qualitative risk register | Reduced number of risks above threshold |
| Linked to value drivers | | |
| Minimise disruption | Hours lost per week | $-b\%$ p.a. |
| Enhance quality of life for residents | Number of recorded defects | $-c\%$ p.m. |
| Conduct and behaviour | | |
| Attitudes to problem solving | Number of problems resolved within time | $+d\%$ p.a. |
| Communications | Number of non-conformances in document control system | $-e\%$ p.a. |

The typical objectives should be project specific and may be classified under headings similar to those indicated in Table 9.3 (based upon a partnering workshop for a social housing development).

## *Organisation and communication chart*

One of the keys to successful partnering is getting people to communicate effectively with one another. As a starting point it is useful to identify the individuals at different management levels in each of the partnering organisations. Encouraging people in the different organisations to communicate with their peers reduces communication barriers.

■ Get each organisation to provide their project organisation chart and identify, between organisations, who communicates directly with whom (named individuals).

■ People generally communicate directly with their counterparts at the same management 'level' within organisations.

■ Establish the 'levels' at which decisions are made.

■ Draw a combined chart of the output (e.g. Figure 9.2).

| Organisation level | Partner organisation A | Partner organisation B | Partner organisation C | Partner organisation D | |
|---|---|---|---|---|---|
| Strategic | ← - - - - | - - - - - - - - | - - - - - - - - | - - - - → | |
| Programme | ← - - - - | - - - - - - - - | - - - - - - - - | - - - - → | Escalation |
| Project | ← - - - - | - - - - - - - - | - - - - - - - - | - - - - → | |
| Operational (site) | ← - - - - | - - - - - - - - | - - - - - - - - | - - - - → | |
| Tradesmen | ← - - - - | - - - - - - - - | - - - - - - - - | - - - - → | |

**Figure 9.2** Communication chart

## Establish decision-making procedures

Establishing good communications across the team will, by itself, improve decision-making. Once a project has been launched, much of the decision-making effort is focused on the resolution of potential variations to the master plan, whether these be opportunities or threats. The following procedures relate directly to problem resolution but apply equally to other forms of decision-making. The aim is to establish a simple procedure to resolve problems, or take decisions, with the minimum of fuss and in a collaborative spirit. The team should work on the principal that problems should be resolved at the lowest practical management level in the organisation. Only if this is not possible within an agreed time constraint should the problem be referred to the next highest level, who will have a wider remit of responsibility. The team should draw a chart of the agreed process, using the same management levels as the agreed communication chart. The generic chart should look something like Figure 9.3.

## Problem resolution techniques

Techniques for resolving problems relating to behaviour during workshops are discussed in Chapter 5. One technique for resolving general problems involves calling together a task force. This should be a small group of individuals, with experience in the subject of the problem. They should look at the problem from a wide perspective, for instance, by comparing it to other similar situations or considering how other people or disciplines might resolve it. They should explore the root cause of the problem since this might lead directly to its solution. It is often more constructive

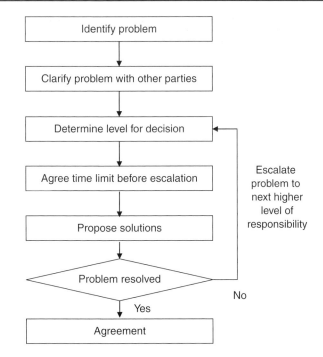

**Figure 9.3** Problem resolution or decision-making procedures

to concentrate on the positive aspects of the problem before addressing its negative aspects. Guidelines for resolving problems should be recorded alongside the procedure exemplified in Figure 9.3.

## *Establish targets and mechanism for continuous performance improvement*

Review the KPIs established earlier and confirm that they are practical and measurable. Agree realistic improvement targets for each and discuss and record how to achieve these. Record the output on flipchart. Note that in the spirit of partnering one partner's problem is everyone's problem. Thus if one partner is having difficulty in performing to the KPI or the improvement targets set, others should help them overcome the difficulty.

For example, one partner could facilitate a value engineering workshop with a supplier to obtain specific improvements in performance or price. Or a client could help a small supplier obtain a better credit rating so he can obtain better discounts (an example of benefit trading).

## Review of issues and observations

During the workshop the study leader will have visibly recorded the issues and observations raised by the partners. He should now check that all of these have been addressed. If not, he should lead a debate with the partners and agree any amendments to previous outputs or actions to resolve. All actions should be owned by an individual and have a completion date.

## Partnering charter

The partnering charter is a statement of the key objectives, values and behaviours that will underpin the partnership (see Figure 9.4). It is not a contract although it should, of course, be consistent with

---

(Project title)

We, the client and members of the supply chain, intend to meet the needs and expectations for this project and to achieve its delivery to the benefits of all parties.

We aim to work together to:

1.    Deliver the project specification to the agreed budget, timetable and standards of quality.
2.    Adopt the constructing excellence codes of practice and guidance where applicable in selecting team members.
3.    Practice team working, trust, respect, fair dealing, effective communication and openness with all in the project.
4.    Provide all necessary skills to develop the project.
5.    Build a balanced workforce.
6.    Seek continuous improvement with appropriate research and innovation to support the project.
7.    Define, manage and present the project with a responsible attitude towards the environment, the local neighbourhood and the health and safety of all.
8.    Inform everyone involved in the project of these commitments as well as other key people within our organisations.
9.    Monitor performance and provide feedback to all parties during and after the project.
10.   ...(Other agreed commitments).

**Signed**   (*by all partners*)

---

**Figure 9.4** An example of a partnering charter

any contract(s) into which the partners have entered. It is a series of statements of intent and behaviour that summarise what has been agreed before and during the workshop. It is normal for all partners to sign a printed copy of the charter at or soon after the end of the workshop.

## 9.12 Mind mapping

Mind Maps® were first developed by Tony Buzan in the 1960s. They provide a simple graphical way to display information in a manner that, in his words, 'unlocks the potential of the brain'. In the context of value and risk management, mind maps enable a user to three things:

■   To capture the information given during one or more presentations on a subject and group related themes together in a way which is easy to read at a glance.

■   To prepare to give a presentation on a subject that is easy to recall and requires a bare minimum of reference during the presentation. This enables the speaker to present with confidence with a minimum of preparation.

■   To gather and display large amounts of information on a single sheet of paper, showing the relationships between the different pieces. This can be used as an alternative presentation form for a function analysis or to display a synopsis of the issues to be addressed during a study.

The basis of the mind map is to use a minimum of key words or images joined together by lines to convey relationships at a glance. Use of different colours can highlight related themes. The method, illustrated in Figure 9.5, is to start at the centre of the piece of paper with the core theme. Then radiate outwards (usually in a clockwise direction) with related topics. Each 'level' away from the centre represents another tier of detail in the description. For further reference the reader is referred to Tony Buzan's website www.mindmap.com.

## 9.13 The Delphi technique

This simple technique helps groups to converge from a variety of viewpoints to a consensus agreement fairly rapidly. The study leader first asks members of the group to explain the situation and give reasons for their particular point of view. The question might,

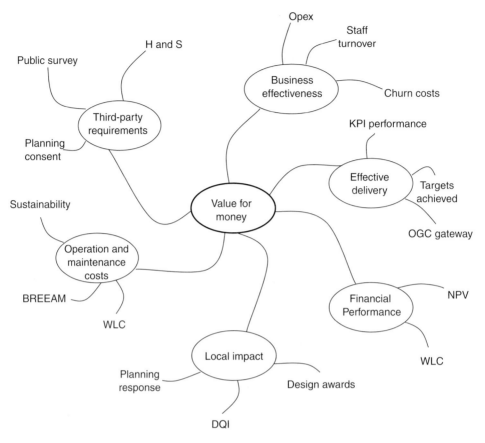

**Figure 9.5** Mind Map describing the key value drivers and possible metrics

for example, relate to the ideal size of a particular type of restaurant, stated in terms of the numbers it could accommodate (or covers). He would then ask all members to write their individual option size on a piece of paper, without sharing this information with their colleagues.

He then would analyse the results in graphical form to show how many people had selected an option size within, say 3 or 4 ranges of size, covering the entire spectrum of returns. He shows this analysis to the group and asks people within each group to explain why they had selected their preferred size.

When all have had their say he will ask them to consider all the arguments that have been put forward and invite them to write down their revised estimates. Usually, after a few such cycles the group will converge on a common view that they can all support.

# 10 Value management techniques

## Summary

- This chapter describes some of the techniques used most commonly, but not exclusively, in value management.

- It opens with a description of the evolution of function analysis and the function analysis system technique (FAST).

- It explains different types of FAST diagrams and how these relate to the value cascade.

- It explains the principles of cost/worth and of improving value through multifunctionality.

- It introduces value trees and the simple multi-attribute rating technique (SMART).

- It then describes value profiling or value benchmarking and options selection.

- It describes several techniques for weighting a multiplicity of items.

- The fundamental techniques of generating innovative ideas (creativity) is described, followed by a section on their evaluation, the development of value-adding proposals and how these will be implemented.

- Other techniques described in this chapter include the development of scenarios, the principles of target costing and function performance specification.

## 10.1 Value management techniques

The techniques described below are commonly, but not exclusively, used in the conduct of value management studies.

## 10.2 Function analysis

Function analysis provides one of the defining characteristics of value management and differentiates it from many other problem-solving techniques. For many people there is a paradigm shift in their understanding of the nature of a project when they make the transformation to describing what things do (the function approach) from describing what they are (the elemental approach). It enhances their understanding of the project, highlights areas of risk and transforms their abilities to generate innovative solutions to problems that hitherto may have eluded them. Function analysis is a key tool leading to a better understanding of risk in a project. The greatest risk of all is that the project fails to deliver the functionality (or benefits) expected of it.

There are many approaches to function analysis, some very structured such as the function analysis system technique (FAST) which is favoured by many practitioners and practiced widely in the United States, to less formal methods such as value trees which can provide an equally rigorous functional model of the project.

In this section we will describe some of the common approaches to function analysis. There is no *right* way to do it. The study leader should select the approach that best suits the project team. There are many who claim that value management without a full understanding of function is not effective value management. The same should be said for risk management.

One of the core principles of value management, introduced in Section 2.4, is that it focuses on achieving successful outcomes more than the process of getting there. Function analysis provides a very powerful tool to identify intended outcomes. In developing a function model, the study team are forced to make a very clear definition of the project by answering such questions as:

- What are we trying to achieve?

- What must we get right if we are to achieve it?

- What considerations do we need to bear in mind while designing it?

- How do various design solutions contribute towards achieving the desired outcome?

The function analysis of a component or of a process forces a similar level of understanding of the component or the process.

The method can therefore be used when studying components and processes, not just on projects.

Function analysis is not difficult but people who are used to thinking elementally, sometimes have difficulty in thinking functionally. Wrapping the process up in jargon is unhelpful. Study leaders should consider expressing function analysis in terms with which the team can readily identify. This does not mean that they should take short cuts or be less than rigorous. It is simply that the terms used sometimes get in the way of the process.

## Allocating cost to function

When he first developed value analysis in the 1950s, Miles established one of the underlying principles of value analysis by dissecting products into functions. The method he developed was one of random function determination in which functions are not linked in any logical manner. The method provides the ability to develop a function cost matrix in which the costs of performing each of the identified functions maybe determined by allocating the elemental costs across the functions. See Table 10.1.

## Verb–noun protocol

He also developed a concise and accurate method of describing each function by way of two words: an active verb and a measurable noun.

Use of these two words provides a way of concisely describing the function to be performed without stating the solution. By

**Table 10.1** Function cost matrix

| Element | Cost | A | B | C | D | E | F | G | H |
|---|---|---|---|---|---|---|---|---|---|
| 1 | 100 | | | | | 100 | | | |
| 2 | 250 | 50 | 100 | | 50 | | 50 | | |
| 3 | 30 | | 30 | | | | | | |
| 4 | 50 | | | 50 | | | | | |
| 5 | 600 | 150 | 150 | | | 50 | 100 | 50 | 100 |
| 6 | 25 | | | | 25 | | | | |
| 7 | 450 | 320 | | 50 | | | | | 80 |
| 8 | 275 | | | 60 | | 120 | | | 95 |
| 9 | 730 | 600 | | | | | | 130 | |
| **Totals** | **2510** | **1120** | **280** | **160** | **75** | **270** | **150** | **170** | **275** |

**Table 10.2** Active verbs and measurable nouns to describe functions

| Active verb | Qualifiers | Measurable noun |
| --- | --- | --- |
| Maximise | World class | Environment |
| Minimise | Best in class | Image |
| Satisfy | Consistent | Performance |
| Create | Conforming | Facility |
| Display | Abnormal | Requirements |
| Enable | Third party | Energy |
| Prevent | Legal | Ambience |
| Enhance | Corporate | Productivity |
| Diminish | Working | Comfort |

expressing problems in this manner one can raise the debate to a much higher level of abstraction and thus encourage greater innovation.

Although purists might disagree, some find the two-word format a bit restricting. Qualifying phrases or adjectives are helpful. This is especially useful when using function analysis to define a project brief. The difference between 'create facility' and 'create a world-class facility' or 'create the best facility in London' can make all the difference. Table 10.2, therefore, includes some commonly used adjectives.

## 10.3 Function analysis system technique

In 1968 Charles Blythway developed function analysis into a comprehensive technique for logically linking functions. He christened this the function analysis system technique (FAST). This method and the term FAST diagram are commonly used by most practitioners of value management when building their functional models.

Readers should note that FAST is not the only technique for doing this and we describe other approaches later in this chapter.

What the FAST process does is to permit people from different technical backgrounds to use a common language to describe and link the functions of complex systems, using non-technical language. In building the FAST diagram the team will interact and communicate with one another effectively to arrive at a logical diagram that they can all understand and agree with. A very simple FAST diagram for part of an office building is illustrated in Figure 10.1.

**How?** ⟹

⟸ **Why?**

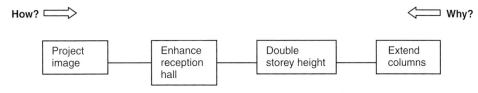

**Figure 10.1** Simple FAST diagram for part of an office building

Let us explore the language and the method of building a FAST diagram in more detail.

## How? and why?

By asking *how*? we move from left to right within the diagram and from higher abstraction to lower abstraction. (The term abstraction is used because, while one can touch and feel a physical element such as a column, this is not possible with an image, which is an abstract concept.) By asking the question *why*? we move from right to left in the diagram and from lower abstraction to higher abstraction. We can test the logic of the links we create by asking the questions how? and why?. If by asking how? of a function of higher abstraction it does not lead to the function of lower abstraction, then there is an error in the logic and we need to think again. This is illustrated in the very simple FAST diagram in Figure 10.1.

This 'how/why' logic provides the key to developing a logic linked function diagram. Essentially randomly generated functions (as used by Miles) can be logically linked through the use of the questions 'how' and 'why'.

## Getting to the heart of a problem

Let us consider the situation in a hospital ward where there is a building project in progress. Assume that there is problem arising due to dust and dirt from the building operation accumulating in the ward faster than it can be cleaned up.

The manager responsible for maintaining the cleanliness of the ward might state this problem by saying

> There is an awful lot of dust and dirt coming in. The cleaners try to keep up but morale is suffering as more and more just keeps on coming in. Some of the patients are complaining and are very nervous

that they might get an infection. Life is bad enough in the ward without these worries. It was particularly bad last week when they were working on the air conditioning system. They don't seem to care. We will have to terminate the building contract if this does not get better.

Note the way in which the speaker in this example described the *symptom* and not the root cause of the problem. His 'logic' leads him to a pretty drastic solution. The action of terminating the contract would probably solve his problem but would create more problems for others. This is not a satisfactory solution.

What is needed is a way to express this problem in a more analytical manner. The main problem in the above example is that there is perceived risk of getting an infection from the accumulation of dust and dirt. There may be other subproblems which can be separately identified. Focusing on the main problem as expressed above, one of the solutions could be to improve the cleaning. By asking the question *how* do we improve the cleaning, we will come up with some solutions, for example: use more cleaners, use power cleaning machines, use antistatic wipes and so on.

By asking the question *why* do we wish to improve the cleaning, one answer might be to 'eliminate dirt'. If we now ask the question *how* do we 'eliminate dirt' we broaden the potential solutions. We could eliminate dirt by the functions: improve cleaning (see above), seal the windows, pressurise the ward, eliminate the source of dirt and so on. By raising the discussion by just one level of abstraction we have generated some much more radical solutions, some of which are probably much more practical.

If we ask *why* eliminate the dirt? The answer might be to reduce risk of infection. If we now ask *how* do we reduce the risk of infection we might come up with the following potential solutions: eliminate dirt (see above), vaccinate patients, sterilise the ward, test dirt and so on. Once again by raising the level of abstraction we have broadened the quest for a radical solution. The diagram in Figure 10.2 shows a distinct logic link between these functions which we have identified.

Note that it is accepted practice when developing a FAST diagram to draw functions of highest abstraction on the left and lowest abstraction on the right. The direction of travel (from higher to lower abstraction by asking *how* and from lower to higher abstraction by asking *why* is indicated by the arrows at the top of the diagram in Figure 10.2).

How? ⟹                                                                    ⟸ Why?

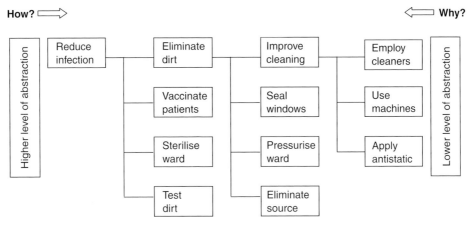

**Figure 10.2** FAST diagram showing differing levels of abstraction

> We can immediately see that this very simple logic provides a powerful tool in the quest for innovative solutions to a problem.

## Dependencies

> In these diagrams, a function of lesser abstraction is dependant upon functions that are more abstract. This is termed dependency. The lower order functions (those on the right-hand side of the diagram) are dependant upon the higher-level abstraction functions, those to their left. If we change a function of higher-level abstraction, the functions to the right (the lower level of abstraction) will change. This is because, as we saw in the example of the hospital ward, the functions to the right are effectively solutions to the function on the left. Thus changing the higher-order functions effectively changes the product. For example, in the example illustrated in Figure 10.3, the original design solution for the project image was to enhance the entrance hall by doubling its height by extending the columns in this area. Another way to enhance reception could have been to create an atrium. How? Erect steel structure (there are no longer any columns to extend).

## Innovation

> When using a function diagram to generate ideas for improvement (see below), there is far more scope for innovation if we address functions of high-level abstraction. Addressing lower order functions will effectively be looking at just one response to the design brief and provide less scope for fundamental change.

How? ⟹                                                                    ⟸ Why?

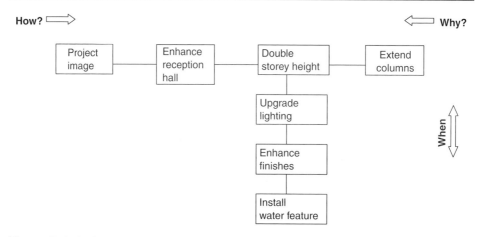

**Figure 10.3** FAST diagram, the 'when' dimension

## Basic functions

For most products there is one overriding function for which that product is intended. This is called a basic function. It can never change unless the product changes. For example, the basic function of a corkscrew is to draw corks. If the corkscrew is not suitable for drawing corks it is useless. Corkscrews have been the subject of more ingenuity than most simple household objects. They have generated many imaginative enhancements, all of which are aimed at making them more efficient in their basic function. But unless they can perform the basic function 'draw corks' they will not be of any use. There maybe other ways to draw corks. By taking the function to a higher level of abstraction 'open bottle' we can begin to explore these. For example, there are devices to pump the cork out by pressurising the air inside the bottle. The basic function of that instrument is to draw corks or remove cork but it is not a corkscrew. A screw top bottle provides the means to open bottle. Its function is one level more abstract than drawing cork. It does not describe a corkscrew.

## Lower order functions

The basic function in a FAST diagram for a building is the primary purpose for which that building is constructed. This may be expressed as 'meet project objective'. In order to achieve the project objective it is necessary for the project team to deliver a number of other functions. These are called primary functions. These primary functions amplify the description of the project objective in terms of what the customer is expecting from the facility. Each of these primary functions can itself be broken down into a further subset of dependent functions, secondary functions, which are likely to describe those things that the project delivery team, and in particular the design team, need to bear in mind when involving potential design solutions. Each of the secondary functions can then be broken down in similar way to tertiary functions and so on, providing ever more detail in describing the requirements of the facility. The lowest level of abstraction in this kind of diagram is likely to be the elements which are proposed by the design team. Depending upon the level of detail into which the study must go (the right-hand scope line) these elements could be functional areas of the building or the components from which it is constructed. In some cases it may be necessary to break each element down onto its component functions.

## *Scope lines*

It may be that we wish to limit the scope of the problem under question. We can do this by introducing scope lines as indicated in the FAST diagram in Figure 10.3. This now describes the limits of the problem under discussion.

## *Other FAST terms*

### When

Sometimes we wish to identify other things which occur simultaneously with an identified function. The FAST diagram uses the term 'when' to describe this situation. The 'when' axis runs vertically. Whereas the functions linked horizontally through 'how' and 'why' are dependant functions, the functions indicated on the when path are independent (support functions). Independent functions describe an enhancement or a control of the function from which it stems. They do not depend upon another function of higher level of abstraction. Independent functions are located above the logic path of the functions to which they are secondary. Moving downward on the when path indicates an activity that will be triggered when the dependant function happens. This logic is indicated in the diagram in Figure 10.3.

### And

There are two other useful terms used within the FAST diagram. The first of these is 'and' to indicate when two dependant functions happen simultaneously. Their importance can be indicated by

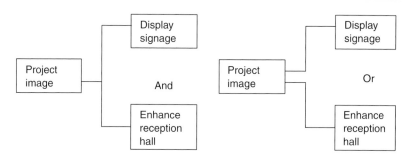

**Figure 10.4** FAST diagrams, 'and', 'or' branches

positioning them above or below the main dependant function logic path.

**Or**

The other key word is 'or' to describe that eventually either one dependant function will happen or another depending on circumstances. The importance of these functions relative to one another is signified by their positioning above or below the main logic part of the dependant functions. The use of 'and' and 'or' are shown in the FAST diagrams in Figure 10.4.

For most building applications the above rules are sufficient to describe a building in the level of detail required for value and risk management. Enthusiasts of FAST modelling may, however, wish to develop more complex models. They should refer to specialist reading on the subject.

## Numbering functions

By numbering the functions in a logical sequence in a FAST diagram it is possible to relate functions to other items. For example, one can draw up a matrix such as RASI diagram (R = who is responsible, A = who must approve, S = who should support, I = who need to be informed). This is a useful way of indicating who is responsible for undertaking actions inferred by the functions and to show how many people will be involved in the activity being described. Figure 10.5 shows a simple FAST diagram for a building, for example, an office building, and a RASI diagram indicating responsibilities.

Numbering the functions is good discipline for identifying them for any later analysis. A further extension of relating to other activities with the FAST diagram is to be directly linked with the risk register. We saw that one of the greatest risks in any building

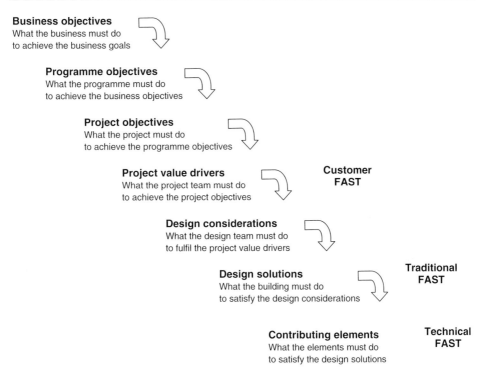

**Business objectives**
What the business must do
to achieve the business goals

**Programme objectives**
What the programme must do
to achieve the business objectives

**Project objectives**
What the project must do
to achieve the programme objectives

**Project value drivers**
What the project team must do
to achieve the project objectives

**Customer FAST**

**Design considerations**
What the design team must do
to fulfil the project value drivers

**Design solutions**
What the building must do
to satisfy the design considerations

**Traditional FAST**

**Contributing elements**
What the elements must do
to satisfy the design solutions

**Technical FAST**

**Figure 10.5** Relationship between FAST diagram types and value cascade

project is that the resulting building fails to deliver the functionality expected of it in full. Clearly, by expressing the functionality in the form of a FAST diagram (or any other type of functional model) we create a structure for identifying risks and uncertainties against each of the functions. We will explore this in more detail later in this chapter.

## Types of FAST diagrams

The evolution of FAST stems from the application of value analysis to product improvement, mainly confined to manufacturing. With the evolution of value analysis through value engineering to value management, applied to a much broader range of subjects, different types of FAST diagrams have emerged.

### Traditional FAST

This is the original form of function model developed to describe what the component parts of a product must do. It may be applied at any level, system, subsystem or component level.

## Technical FAST

This applies mainly to products and goes into great detail in describing its functionality. It is very similar to traditional FAST. It is used widely in manufacturing applications. It is generally used at subsystem or component levels.

## Customer FAST

With the evolution of modern management thinking and the realisation that value is determined by the customer, not the producer or supplier, a form of FAST has evolved which focuses on customers requirements. Here the basic function represents what the customer expects to get from the item under scrutiny. In applying value management in the early stages of a building project, it is this type of FAST diagram which is most useful. It is the most abstract of the three types of FAST diagrams described above.

While the method of constructing a FAST diagram is identical to that described earlier in this chapter, the left-hand scope line (the highest level of abstraction under consideration) will be placed just to the right of the basic function of the building.

## *FAST and process flow diagrams*

Those who are used to creating process flow diagrams or flowcharts may find this 'how/why' logic somewhat confusing. They might think that the diagram is the wrong way round. The reason for this is that a process flow diagram starts at the beginning of the system and ends with what it is trying to achieve. The FAST model on the other hand starts with the goal and ends with the beginning of the system. (The goal being the higher level of abstraction and the beginning of the system being the lower level of abstraction.)

## 10.4 Cost/worth

A key principle of value analysis is to analyse each of the functions and assess what it actually costs to perform the function, using the functional cost matrix described in Section 10.2. By comparison, the team assesses the lowest cost at which the function can be performed, referred to as the function's 'worth'. A comparison of the two figures gives an indication of the 'cost/worth' of the function being examined. The functions whose cost significantly exceeds their worth (being defined as the lowest cost at which that function can be performed) may warrant further study to explore whether they can be performed in a different way at less cost.

For example, the basic function of the granite paving in the Broadgate Development alongside Liverpool Street Station in London might be to 'support pedestrians'. The cost of the granite paving is likely to be in the order of say £50 per square metre. The lowest cost at which one could perform the same basic function, support pedestrians, is to use concrete paving slabs at a cost of, say, £5 per square metre. The cost worth ratio therefore is 50 to 5 or 10 to 1. Value engineers may consider that there could be other ways to perform this function at lower cost.

## But ...

While this approach may be valid when looking at the manufacture or assembly of systems, it is dangerous when exploring how each of the components contribute to the function of a whole building. The analysis ignores the other functions to which the granite paving contributes, not least of which is the aesthetics of the area. The function 'project appropriate image' is performed by the granite but is unlikely to be satisfied by the concrete paviors. Similarly, granite performs an additional function 'resist wear' considerably better than its concrete equivalent. There are also other functions that granite satisfies and which the concrete pavior does not.

## Need to identify the full functionality

It is vital, therefore, that in considering the basic function of a component, the other functions to which it contributes are taken into account. This can only be achieved by drawing up a full logic linked function diagram to which costs are applied. In this case, the granite paving slab contributes to more than one function. This is common to many components. *All* functions to which the component contributes must be taken into account when assessing its 'worth'.

## Cost cutting

Thus the simple concept of cost/worth can be flawed unless great care is taken to explore all the functions performed by the component under examination. It is a common criticism of value engineering that the associated functions to which a component contributes are ignored, resulting in cost cutting in which functionality is lost. The resulting building either fails to match up to the expectations of the users or, worse still, undermines the whole concept of its design. Such practice is clearly detrimental, destroys value and must be avoided.

## 10.5 Multifunctionality

The concept of multifunctionality is an excellent way to add value. If a single space within a building can perform many functions, it saves having to build additional space for each function performed. Modern schools play an increasing part in acting as community centres. Because the curriculum is governed by a timetable, it is possible for the same space to serve many functions. The dining area, for example, is only used for meals for a small part of the day. It may double as a teaching space during the day and serve as a community centre in the evenings and weekends.

A splendid example of multiple functionality was to be found in Isambard K. Brunel's final achievement, when constructing the largest ship in the world in its day, the Great Eastern (see Figure 10.6). Because this was such a radical departure from traditional design and a step change in size from any ship that had previously been constructed, Brunel applied his experience from building bridges. He designed the hull as a giant girder which was capable of spanning between large waves while the ship was at sea. One of the key features of his design was the use of a cavity hull, that is to say, two skins to the hull of the ship connected by sheer plates, so that they acted as a single sandwich construction.

**Figure 10.6** The Great Eastern – an example of multifunctionality

Not only did this provide great strength but in the event of damage to the hull it provided a second line of waterproof resistance.

If the Great Eastern had struck an iceberg, like the Titanic, it is unlikely that it would have sunk, since the inner skin of the hull would have prevented the water from flooding the ship. The twin hull performed a third function which was to minimise weight. Minimising the weight of a ship of this size was crucial to maintaining its sea worthiness and, indeed, enabling it to float.

## *Eliminating redundancy*

The function diagram is a powerful way of demonstrating the contribution made by each and every component to the completed building. It is immediately apparent that if a component makes no discernable contribution to the required functionality of the building or is redundant, it can safely be omitted.

An example of this occurred when value engineering a prison in the United States. The design included two reception areas, one for the public, the other for staff. There was a strict requirement to preserve unimpeded sight lines in all parts of the reception area. It was therefore necessary to install two security guard positions. Both performed the same function, preserve sight lines. A simple rearrangement of the reception area eliminated the need for the second security guard's position and the need for the second guard. Not only was the function 'preserve sight lines' maintained but the proposal generated significant savings. The first saving arose from not having to build a second security position. The second saving came from not having to man the position. Manning a single position required the employment of no less than five individuals (to provide cover for 24 h a day and allow appropriate rotas for breaks, holidays, etc.). The latter saving, on a whole-life cost basis, was by far the larger. No functionality (benefit) was lost and significant cost was saved in both investment and operations. This represented a significant increase in value and is in stark contrast to cost cutting.

## 10.6 SMART methodology

Recognising this weakness in the cost/worth principle of early value engineering, Professor Stuart Green of Reading University evolved his SMART (simple multi-attribute rating technique) approach to value management in the mid-1990s. This introduced two evolutionary concepts from FAST methodology. The first was the value tree. Instead of expressing the essential components necessary to fulfil the project objectives by means of simple two-word functions, the SMART methodology uses the concept of a value tree to link functions. The second innovation was to create an importance hierarchy for the functions.

## The value tree

The value tree is similar to the FAST diagram in that it, normally, begins on the left-hand side with a statement of the project objectives. The answer to the question 'How?' is expressed in simple value-adding attributes required to deliver the project objective. Each of these is broken down into the attributes that add value to that branch, thus building a tree of decreasing abstraction from left to right. This is illustrated in Figure 10.7 for a laboratory.

## Importance hierarchy

The client ranks the value-adding attributes in order of their importance in achieving the project objectives. Under the SMART value tree the weighting for each of the values is expressed as a decimal

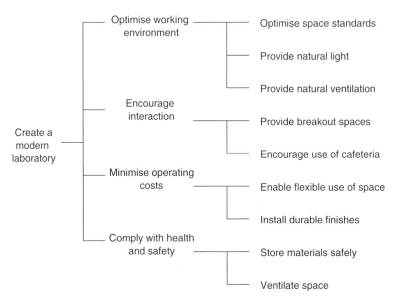

**Figure 10.7** A SMART value tree for a laboratory

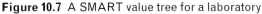

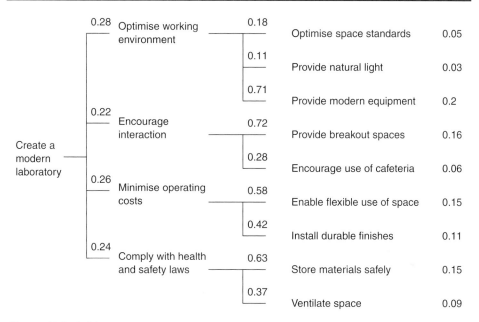

**Figure 10.8** A SMART value tree for a laboratory with importance weights

of less than 1 such that for any level of abstraction for any single attribute, the total weighting adds to one. This is illustrated in the second diagram of the same laboratory in Figure 10.8.

This enables the value management team to do two things. By attributing costs to the value tree (by distributing the cost of individual elements across the branches of the tree to which they contribute), it is possible to assess the relative costs to undertake a function with its importance in the overall project. If lots of money, time or other resource is being spent on a function which is of little importance, it is an obvious target for the question 'how else can we achieve this function for less resource?'

## 10.7 Value drivers

Building professionals, particularly senior management, often find it difficult to come to terms with the rigour of function analysis to describe their building. This is in part due to the language used and the word function. People generally infer some physical function undertaken by a product, for example, a handle *turns* something. By contrast, in the analysis of a building, many of the functions are not physical in nature. Those functions which describe aesthetics or human behaviour may be quite abstract. Since these abstract functions are essential to adequately describe a building, people find the concept of function analysis somewhat alien. Using the term value driver, in place of primary function, overcomes this difficulty. Value drivers are those things which contribute to the value of the building and are readily understood.

The value drivers shown in Table 10.3 were derived from work with the Design and Build Forum (now since it merged with Reading Construction Forum). These can be used to describe the required functionality (in its purest sense) of any building. For any building it will be necessary to satisfy all value drivers. In addition, the relative importance of the value drivers provides a very clear description of the project priorities. Table 10.3 lists these generic value drivers (in no order of importance) and provides an explanation for each (this is a replica of Table 6.4 and is duplicated here for ease of reference).

**Table 10.3** Generic value drivers for buildings

| | Value driver | Addresses the question |
|---|---|---|
| 1 | Enhance/achieve desired financial performance (of the building) | Is the building affordable? |
| 2 | Manage the procurement process effectively (maximise project delivery efficiency, minimise waste) | Are the project management processes efficient? Are the right people engaged? Is the supply chain effectively managed? Are materials and labour used efficiently? |
| 3 | Maximise operational efficiency, minimise operational costs | Does the building work well for the end users? |
| 4 | Attract and retain employees/occupants | Is it a nice place to work? |
| 5 | Project the appropriate image | Does the building convey the appropriate image? |
| 6 | Minimise maintenance costs | Is the building easy to maintain and clean? |
| 7 | Enhance the environment | Is the building environmentally friendly? Is it built using the principles of environmental sustainability? |
| 8 | Comply with third-party constraints | Does the building conform to legal and other external stakeholder requirements? |
| 9 | Ensure health and safety during project implementation and in operation | Is the building safe to build and use? |

Development of generic value drivers offers an advance on the generation of random value drivers for each and every project. While random value drivers may be just as effective in getting the team to describe the functionality of a building and to express the requirements of that building in terms that everyone can understand, the resulting function analysis is unique to the circumstances under which it was derived. This precludes the ability of comparing one building of similar type with another. The use of the generic value drivers opens the way to build comparisons between different buildings whose objectives are similar.

## Saving time

The development of a set of building-type specific value drivers based upon the generic value drivers described above enables a team to express the required functionality of the building they require in far less time and with less difficulty than if they had started from a blank sheet of paper. Since the value management team are under increasing pressure to conduct the function analysis in less and less time (in the early 1990s it was not uncommon for a value engineering workshop to extend over 4 or even 5 days whereas by the early 2000s workshop durations have reduced to 1–2 days maximum) the ability to build a rigorous function model and communicate the client's intent to the project team in a short space of time is crucial.

An alternative temptation, which we strongly discourage, is to omit function analysis altogether. This is to omit one of the cornerstones of the process.

Using the generic value drivers, or the standard value driver model for the type of building under study, the study leader can often build an outline value driver model (or function analysis) in advance of a workshop, from the information gathered during the briefing process. He can circulate this to the value management team in advance of the workshop and then explore it in greater detail, involving the whole workshop team, during the workshop. In this way, the learning process provided by building the model with the team is retained.

## Clarity of intent

The use of the language in describing each of the functions in the value driver model is crucial in specifying the quality of the facility which is expected. For example, 'world class' means that, at the time of construction, you will have to scour the world to find a facility of comparable quality. For a sports facility, 'international competition class', is likely to be far less onerous in the use of resources than a facility which is deemed to be 'world class'. The requirement to provide for the enhanced functionality to raise the quality from 'suitable for international competition' to 'world class' may not be required by the client or be affordable. This distinction in language gives clear direction to the design team and does not raise expectations unnecessarily. The reduced functionality is adequate to satisfy expectations and avoids wasting resource to achieve unnecessary additional functionality. It represents better value for money.

The client body promoting a proposed major sports complex had high aspirations. They were using words like 'world class' to describe the facilities that they wanted. At the project definition workshop the team developed the function diagram and identified the following value drivers:

1.  Deliver appropriate facilities which should be of international flagship quality and innovative.

2.  Create Landmark design to a high standard, reflects best practice models and delivers value for money.

3.  Ensure value for money supported by robust financial and commercial conditions balancing revenue between the client and the operators.

4.  Ensure timetable compliance to meet key milestones.

5.  Ensure compliance with legal and statutory constraints.

6.  Foster stakeholder support by building the right perceptions in the project.

It soon became apparent that to achieve 'world class' facilities would be unaffordable as well as unnecessary. The function tree was developed to facility level, providing the opportunity to link the project objectives, through the value drivers to the design considerations associated with each facility. Careful choice of the language used to describe each design considerations provided guidance to the design team to design facilities that were likely to be affordable.

Not only did the use of the value driver model (in this case a clear example of a value cascade) clearly articulate the client's intent but it proved to be an excellent way to brief the project team in the early stages of the project. This was particularly valuable for the recently appointed new members to the team. Not only did it provide a very effective briefing tool, but it provided the opportunity for new team members to contribute ways of improving the value of the proposed scheme.

## 10.8 Value profiling (or value benchmarking)

In Section 2.11, we saw how value drivers (primary functions), weighted to reflect their relative importance, may be used to provide a measure of project value.

The same technique provides a powerful tool to describe the client's value priorities, or his value profile. This, in turn, opens the way to identifying in those parts of a project which provide most potential for adding value and how well the project is developing. Value profiling takes place at the outset of a study during the pre-workshop briefing stage. The study leader first identifies the value drivers with the team. He then asks the client body (not his advisers or consultants but including the end users) to weight the value drivers in terms of importance to his business. The value team then identifies objective metrics for each value driver and agrees with the client acceptable ranges for each, from unacceptable to delight. The team then assess where the project now stands within each of the ranges of a scale of 1 (=unacceptable) to 10 (=delight).

They may also identify targets within each range for the study to achieve. The resulting table provides a clear indication of the relative importance of each value driver, a picture of current performance against value, the deficit for each between the target value and current performance and a measure of the current project value.

This principle is illustrated in Figure 10.9 which shows the value drivers for an airport terminal. The current value profile indicates a value index of just 360 against a benchmark target of 750. Clearly there is room for improvement. The table also shows that focusing efforts on the value drivers 'encourage shopping' and 'complete on time' will be most rewarding, since it is here that the 'value deficit' (the difference between the target values and actual) are greatest.

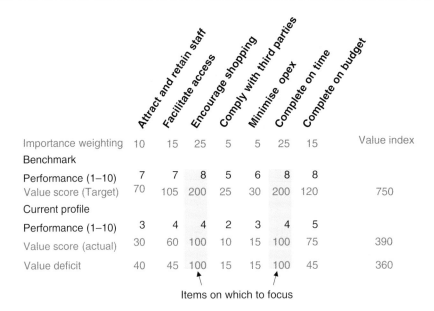

| | Attract and retain staff | Facilitate access | Encourage shopping | Comply with third parties | Minimise opex | Complete on time | Complete on budget | Value index |
|---|---|---|---|---|---|---|---|---|
| Importance weighting | 10 | 15 | 25 | 5 | 5 | 25 | 15 | |
| Benchmark | | | | | | | | |
| Performance (1–10) | 7 | 7 | 8 | 5 | 6 | 8 | 8 | |
| Value score (Target) | 70 | 105 | 200 | 25 | 30 | 200 | 120 | 750 |
| Current profile | | | | | | | | |
| Performance (1–10) | 3 | 4 | 4 | 2 | 3 | 4 | 5 | |
| Value score (actual) | 30 | 60 | 100 | 10 | 15 | 100 | 75 | 390 |
| Value deficit | 40 | 45 | 100 | 15 | 15 | 100 | 45 | 360 |

Items on which to focus

**Figure 10.9** Value profiling (or value benchmarking)

## 10.9 Option selection

The same weighted value driver model provides an objective way of making decisions. By using an option evaluation matrix (originally developed as the combinex method by Carlos Fallon) the team can assess the relative benefits of each option using the weighted value drivers as evaluation criteria. Clearly, the best option will be that which satisfies the value drivers (the very reasons for carrying out the project) best.

By multiplying the weighting of each value driver by the degree to which the options satisfy them (on the scale of 1 – poor, to 4 – excellent) the team arrives at a value score for each option. It is better to use a 4-point scale for this assessment than a 5-point scale which is commonly used. The reason for this is that there is a tendency for groups to pick the midpoint number 3 on the 5-point scale. The 4-point scale forces them to select above or below the mean by choosing 2 or 3. Adding all the value scores across all the functions for each option gives a total value score for that design option. The option with the highest value index is that which best satisfies the client's requirements in the project objectives. An example of an option evaluation matrix is given in the Chapter 12.11.

## *Value for money*

In a further sophistication of this technique, dividing the value score by the total cost of the option yields a value for money index. This will differentiate between two options, each of which has a high value score (i.e. they satisfy the value drivers very well) but one costs significantly more than the other. For example, a Ferrari may satisfy a driver's every need completely, in terms of speed, image and luxury. A Jaguar will probably provide better value for money.

If this type of analysis is to be sound, the way in which the weighting is undertaken should also be scientifically sound. Several techniques for doing this are described in the next section.

## *Sensitivity analysis*

Note, small changes in the weighting of each option against the value drivers can swing the balance between them. Even when using a scientifically sound means for assessing weightings the development of the option selection matrix can be subjective. The same is true when assessing how well each option satisfies a particular function. It is therefore vital that, before decisions are made, sensitivity analysis is performed, varying both the importance weightings and the satisfaction assessments to ensure that the analysis is robust. In the case study outlined above the sensitivity analysis confirmed the choice.

*Note*: Applying such a sensitivity analysis is not code for fiddling the results to get the desired outcome!

## 10.10 Weighting techniques

### *Dots*

Asking the team to allocate dots to their preferred choices is the least scientific but quickest method of indicating the relative importance between a number of items. For a group of, say eight items, each team member is asked to identify their favourites. The study leader should guide the team as to how many favourites to select. The number should be around 20–30% of the items under consideration. They then place dots against each. When all dots have been placed the study leader simply adds up the number of dots against each item. That with the most dots is ranked most important and so on.

The method is very quick but the weightings (numbers of dots) against each criterion is not scientific. They do, however, give an indication of relative importance. This simple method can also be used to make an initial selection of the ideas that the team wish to select for developing into proposals after brainstorming. Any ideas that have no dots against them are supported by no one and may be safely rejected.

## Distribution of points

A more statistically sound way to assign weightings is to ask the team to allocate a fixed number of points, say 100, between the items. Each person can put as many points as they wish against any one item but must use all their points and no more. The study leader notes the allocation by each individual, adds up the total for each. He then divides this by the number of contributors. This gives an average score for each item. The study leader then normalises these scores to arrive at a percentage weighting for each. A suitable proforma for this technique is included in Chapter 11.

It is more interesting for the team if the study leader asks them to imagine that the points that they are allocating are units of their own money. In effect he is then asking the question 'If this was your project, where would you spend the money?'

The method suffers from the disadvantage that the team may not be representative of the whole project team. Some decision-makers may be inadequately represented and be 'out-scored' as a consequence. To overcome this difficulty, the study leader should agree on a fair representation of each of the stakeholders before commencing the exercise. When making decisions that affect the outcome of the project (rather than the processes for delivering it) only the client body, including the end users, should be asked to allocate points.

Sometimes there is significant divergence in individual weightings across the team. In these cases, the average is not representative of all the members' views. The study leader should then apply a convergence tool, such as the Delphi technique to achieve consensus.

## Paired comparisons

A more sophisticated technique (but also more time consuming) is to undertake direct comparisons between each of the attributes

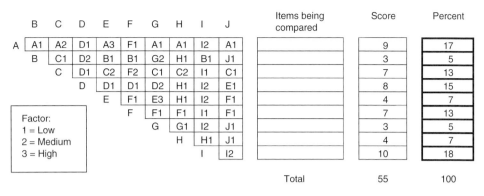

**Figure 10.10** Paired comparisons matrix

being weighted. Figure 10.10 shows a matrix for undertaking this analysis.

In this method, the study leader asks the team to judge each item against just one other, A against B, A against C, and so on until all combinations have been judged. The team may either make a simple comparison by entering just the letter attaching to the 'better' attribute in the box, or they may judge by how much it is better and place the letter plus a number in the box. Scales for assessing by how much one attribute is better than another vary: 1 (low) to 3 (high) being a commonly used scale. The score against each attribute is calculated by adding up the total number of times that attribute appears in the matrix (each entry being multiplied by the scale factor if applicable). The weighting is then calculated by normalising the scores on a percentage basis.

## 10.11 Generating ideas for adding value (creative techniques)

One of the underlying fundamental concepts of value management is to encourage innovative solutions. The value management team can use function analysis to foster innovation.

Traditionally, to improve a design solution or a product, the team will look at the individual components and elements of which the product is made and ask the question 'how can I improve this component to make it better, stronger and cheaper'.

Applying the principles of function analysis and asking the question 'what must the component do?' raises the level of abstraction enabling the team to come up with alternative ways to fulfil the function. This opens the way for significantly enhanced innovation in terms of the solutions.

Imagine that the team wishes to improve an overhead projector. The traditional approach would be to review each of the components and design an improved projector. They might chose to describe its function as 'project image'. They could then ask the question 'how can we improve the way in which we project the image (of the slide) onto the screen?' Various ideas will emerge, some being more practical than others. It is very likely that they will find some way to improve the product. However, it will still be an overhead projector.

If we were to ask '*why* we would wish to project the image (of a transparency) onto a screen?', the answer could be to 'convey information'. This effectively states the problem at a higher level of abstraction than 'project the image' on the screen. If we now ask the question 'how else might we convey information?' we can generate a much broader and innovative choice of solutions, for example:

- Improve overhead projector
- Write on a blackboard
- Use a flipchart
- Video conferencing
- Telephone conferencing
- By e-mail
- Verbal description
- Use of digital projector
- Use an advertisement.

It is clear that the outcome of only one of the above solutions will be an overhead projector. All the others still perform the stated function, 'convey information'. Depending on the circumstances, they may be better or worse than the projector. They can however lead to totally different ways of solving the problem. This is the key to innovation and solving a problem in a way that has not previously been done.

## *Function analysis as the basis for idea generation*

The example above explains the golden rule: always use function analysis as the basis for exploring innovative solutions and new ideas. If value engineering teams omit the function analysis or, even worse, having developed a detailed function analysis for the project lay it to one side, and revert to drawings and elemental cost plans to explore new ideas, they will stifle innovation. The result of such a quest is almost invariably a series of cost or time reduction proposals resulting in reduced functionality which cheapens the scheme. While effective value management may reduce costs, it is certainly not the intent to cheapen the outcome. This implies lower quality and brings the discipline into disrepute. It is essential to maintain or enhance necessary functionality.

In the early stages of the project, design solutions may not be known or costed. It is, however, still possible and valid to build a value driver (function) model. At this stage it may not be possible to attribute costs because of the inability to put a price on fulfilling abstract functions. The team can, however, use the function model as the basis for generating innovative ideas on how to improve the concept. Later in the project, as the design is developed, they can develop the function analysis to lower levels of abstraction and estimate the cost of performing each function using the components and elements which contribute to them. It is still vital to use the function cost model as the basis for generating ideas if essential functionality of the project is not to be lost.

The parts of the function cost model offering greatest potential for value improvement will be informed by the value profile and cost/worth models.

## *Brainstorming*

The most common method of generating ideas in a multidisciplinary value management team is through the technique commonly known as brainstorming. In this method the study leader will invite members of the team to come up with alternative ways of satisfying the function under discussion. He may provide guidelines for generating ideas, for example, by combining two features to perform the same functions (as we saw in the prison example earlier) thus stimulating the team to generate innovative ideas.

The study leader, or members of the team, will record each of the ideas as it is generated, usually on flipcharts or some other similar visible means. Each idea should be recorded visibly so that everyone in the room can see it, using the words of its originator. The ability to see an idea that has been generated earlier may stimulate another idea in the back of the mind of another member of the team. The study leader will continue recording all ideas until the team slows down, indicating that no further ideas are likely to be forthcoming. He will then move to the next function or the next branch of the value tree and again ask the question 'how else can we perform these functions?' He will continue in a similar manner until the entire project or parts of the project under consideration have been comprehensively explored for new ideas.

If a participant wishes to raise an idea relating to a function which has been discussed previously, this should be encouraged not discouraged. The important thing is to capture the idea – never mind that it may be out of sequence. Catching out of sequence ideas is one of the reasons why all ideas are recorded in a visible manner and kept visible at all times. Once the brainstorming session is complete and all areas of functionality have been explored, the study leader may give the team time to come up with any ideas relating to any aspect of the project or issues which has been raised during the various workshop team briefings.

It is good practice to give each idea a unique number. This provides label by which to refer to it.

If it is not practical to cover the entire project in a single session, the team may select only those areas of functionality where there is most scope for improvement, using the concepts of cost/worth and value profile described earlier.

Every workshop generates new and innovative ideas specific to the project under study. These ideas are retained within the records for that particular project but may not be stored in an easily retrievable manner for use in similar circumstances in the future. The concept of value-based thinking demands that ideas from one workshop are easily fed into the creative thought of the next project. Experienced participants will feed ideas from previous projects into a brainstorming session. To encourage truly creative thought the study leader should ensure that:

1.   The idea is relevant to the function under discussion.

2.   It has not been explored before on this project (say in a previous workshop) and discarded.

3.  It is not presented in a manner which stifles original thinking. Tabling previously recorded lists of ideas at the outset of a brainstorming session as one participant's 'contribution' and out of context with the functions under discussion kills any spontaneity or originality. This practice tends to detract attention away from the function under discussion and reduces the whole session to a tabling of 'these are some ideas we've thought of earlier, let's pick the best of these'. It is unlikely to result in innovative solutions or proposals.

## *Overcoming reluctance*

Brainstorming as described above is constrained by the speed at which the ideas can be captured, if they are to be shared in the visible means described above. Certain members of the team may be reticent in voicing their ideas if they feel that their ideas are running counter to the principles of their superiors, or if they are concerned that they may be ridiculed. In these circumstances, some kind of 'confidential' way to generate ideas is to be encouraged.

One method involves each contributor writing their ideas onto slips of paper which they place in a collection point in a convenient position within the room. The advantages of this method are that there is no constraint on the generation of ideas and the speed of gathering the ideas is not constrained by the rate at which the study leader can record them. A disadvantage is that other team members cannot feed off previously generated ideas and there can be significant duplication of the ideas generated. To overcome this, the ideas need to be sorted, after the brainstorming session, into categories relating to functions under discussion and any duplication eliminated. The ideas can then be numbered.

Another method is to place flipcharts on the wall representing each of the branches of the function model and inviting contributors to write, or stick, their contributions onto the flipchart in their own time. While this is not completely confidential, it can be reasonably anonymous if visual aids such as Post-its are used for recording the ideas. This method has the advantage that it is not constrained by the rate at which the study leader can record ideas. With a bit of care each idea can carry a unique number and it is reasonably anonymous. A further extension of this method is the use of large surfaces on which to gather ideas such as the metaboard system in which groups cannot merely record ideas but build these into logic linked scenarios comprising a number of ideas.

This sophisticated type of freewheeling idea generation is useful in stakeholder conferencing in the early stages of a project.

## *Ownership of ideas*

When facilitating any of these idea-generating sessions, it is crucial that the study leader should not impose his ideas upon the team. While it is perfectly acceptable for the facilitator to seed ideas (e.g. from previous experience of similar situations), unless those ideas are seized upon and supported by at least one member within the workshop team, they should not be recorded. One of the requirements of idea generation is that the ideas generated should be owned by the project team. It is the project team who will implement them if they are selected.

## *Defer evaluation*

It is a common failing, during idea generation, to debate merits and demerits of each idea as it arises. The net result of this is to stifle potential original ideas before they are articulated. It also provides the opportunity for the defensive expert to dominate the session by rubbishing everyone else's ideas but his own. The skilled facilitator will ruthlessly stamp out any attempts at judging ideas as they are generated and discourage the overbearing expert from dominating the proceedings. It is easy to demonstrate that, if evaluation is forbidden during generation of ideas, not only will significantly more ideas emerge but, even allowing for duplication, many more will be acceptable.

## 10.12  Selecting the best ideas (evaluation)

A creative team in a good brainstorming session, even in a short workshop, can generate several hundred ideas. Some of these will have greater merit than others. Indeed it is a sign of a truly original free-flowing expression of thought that many of the ideas will be unworkable. By contrast, many of them will form the basis for significant improvement in value in the project. The next stage in a workshop is to select which of these many ideas warrant further study.

The aim at this stage is not to agree to *implement* the idea in question. It is simply to select the best ideas that warrant working up and validating into more detailed proposals. It is only when these proposals are presented to a decision-making body that any decision will be taken as to whether or not to implement them. Emphasising this distinction makes it considerably easier for the team to select ideas of merit.

## Common evaluation criteria

The ideal situation comprises a multidisciplinary value management team containing a balanced mix of characters. One problem with such a team is that each discipline and character group is likely to have their own set of values against which they will judge ideas. Before beginning the selection process therefore the study leader should establish some simple criteria which can be applied by all team members no matter what their discipline or character. The criteria should be linked to the objectives of the study, which are very simple and few in number. A typical set of criteria might be as follows:

- Reduces (whole-life) cost

- Improves time certainty

- Complies with corporate and legal requirements

- Is practical to deliver (within the time and cost constraints).

These simple criteria should be generated by the team, although the study leader might make some suggestions. The study leader should write these criteria up so that they are visible for all to see when making their judgements.

The team is likely to reach rapid consensus on the acceptability of the majority of ideas without explicit referral to the list of evaluation criteria. If they get stuck, the study leader will refer them to the list. This normally resolves the argument.

## Ranking ideas

Using a two-way switch to decide whether an idea should be developed or rejected, with nothing in between, can lead to the rejection of some ideas with potential.

Rather than simply agree on a yes or no to whether an idea warrants further development, it is better to identify two or three categories of ideas and give each of these a number or letter, for example:

- Three could be ascribed to any idea where there is unanimous agreement that it warrants further development into a value engineering proposal.

- Two could be given to an idea that has merit but for one reason or another may not be ideal for development in the short term. It may, for example, require significant testing or validation for which there is insufficient time, or maybe the information required to develop it is simply not available.

- One might be given to ideas which have merit on future projects but are not practical for this particular project.

- Zero might be given to all ideas where there is unanimous agreement that it does not warrant any further development.

## *Shopping lists*

This is not the only way in which ideas may be categorised. Other forms of categorisation could include creating a shopping list of ideas. While all the ideas in the shopping list are feasible, it may not be known, until sometime in the future, whether or not there will be a need to implement them. Creating such a shopping list related to savings and the time within the project lifecycle when they can be implemented could be of great assistance to the project manager when running a project to a tight timescale and budget. If procurement and the project are proceeding well, contingency sums and risk allowances may be relatively untouched allowing the project manager to authorise the implementation of some proposals which enhance the functionality and the benefits to be delivered by the building. Conversely, if things are not proceeding to plan, the project manager can defer or decide against their implementation. Some proposals will affect works which necessarily must take place early on in the project lifecycle (e.g. foundation improvement) and these cannot be reinstated later on in the project. By contrast, some proposals relating to finishes or fit-out can be deferred and implemented much later.

Having agreed on the selection criteria and the categorisation labels for selecting ideas, the study leader must get the team to reach consensus on which ideas should be further developed. In small groups numbering, say a dozen or so, in single workshop session, it will be fairly quick and simple to work through the list of ideas generated during the creative phase and quickly reach consensus on the category of the idea and the lead facilitator will mark up the record accordingly. In larger groups, it may be better to break the team into groups, each reviewing ideas relating to different functions.

A skilled facilitator will, during this stage, exercise control over the overbearing expert, who may try to dominate proceedings, to allow others to voice their opinions.

As a rule of thumb it is likely to take more than twice as long to evaluate a list of ideas than it took to generate them in the first place. During idea generation, the study leader will have stopped any debate on the merits or demerits of the ideas. In this, the judgemental phase, it is essential that the merits and demerits of each idea can be debated in reasonable detail. This takes time.

## Cross-related ideas

Frequently, one idea needs cross-relating to another idea on which a judgement has already been made. This may simply be because it was generated at a different stage during the creative phase but the proposed action is the same. Numbering each idea helps to simplify idea cross-referral (Table 10.4).

## Proposal owners

Once all ideas have been evaluated in the manner described above, it is normal to allocate an owner to each one. He will be responsible for its development into a proposal. This can be done at the same time as selecting an idea for development if the team prefers to do it that way. The owner, who should be an individual rather than an organisation, is then responsible for working with whom so ever he needs to, to develop the proposals in the manner described in the next section. The proposal owner will normally present the outcome to the decision-making panel during the wrap-up meeting. The owner should agree with the team on the timescale by which each proposal requires development.

The above method is ideal for selecting the best proposals from a large number that have been developed in a brainstorming session or other similar idea-generating forum.

A quicker, but less rigorous approach, is for the study leader to invite team members to write their initials (or if anonymity is required, place a coloured dot) beside any ideas that they think warrant development. Clearly any idea that is supported by no one can be rejected out of hand. Those with the most initials or dots against

**Table 10.4** Extract from an evaluated creative idea listing

| No. | Description | Rank | Owner | Cost impact | | | Comment |
|---|---|---|---|---|---|---|---|
| | | | | Saving | Extra | Cross-Ref. | |
| 1 | Defer auditorium | 3 | SB | | | 35 | |
| 2 | Use of reconstituted stone in lieu of natural | See 24 | | | | | |
| 3 | Review standards of facilities | 2 | CM | | | | |
| 4 | Shorter narrower concourse | 0 | | | | | F, unacceptable loss of quality |
| 5 | Shorten street | 3 | CM | | | | |
| 6 | Move entrance | See 5 | | | | 6,20 | |
| 7 | Cut-out duplicate space in laboratories | 3 | CM | | | | |
| 8 | Vary timetable to share space | 0 | | | | | B, adverse impact on timetable |
| 9 | Reduce non-operable space in administration building | 3 | MC | | | 13,50,53,55,56 | |
| 10 | Reduce non-operable space in academic building | 3 | KL | | | 13 | |
| 11 | Move part of scope to PFI | 0 | | | | | D, not practical within timetable |
| 12 | Restructure deal with developer to make economies of scale | 1 | MP | | | | |

*Notes:* Evaluation criteria
  A   Capital and/or revenue cost saving
  B   Impact on core business
  C   Statutory requirements (planning)
  D   Fits short-term programme
  E   No adverse impact elsewhere
  F   Upholds quality.

Key to ranking
3 = develop before wrap-up meeting (within workshop)
2 = develop after wrap-up meeting (within a week)
1 = consider for future development (by end January 2000)
0 = reject.

The words "See…" in the Rank column refer to other ideas (not shown)
The initials in the owner column are the owners initials
The cross references refer to other related ideas in the full listing

them should be debated for selection first, while those with fewer initials or dots can be left until later in the discussion. If time for discussion is short, this method has the merit of evaluating the most popular ideas first. The workshop record will then show those ideas which had the support of members of the workshop team even if the idea itself was not evaluated in any detail.

## 10.13  Developing implementation proposals

During a short workshop there is likely to be insufficient time to develop implementation proposals. Any development is likely to be in outline only. If it is of significant benefit, it will require more work to validate assumptions and work it up to the point where it can inform a decision by senior management. This is particularly true where proposals form part of a broader scenario, built up of a number of proposals, and could result in significant changes to the project. Developing outline proposals during the workshop will ensure that their owners are clear about the nature of the proposals they should develop. This avoids them wasting time developing a proposal which might not be the same as that expected by the workshop team.

It is useful to develop all proposals to a common format. Not only does this assist presentation to management but it also ensures a consistency in the information contained within the proposal.

1.  Summary form: this form summarises the essential information that the management will need to know when considering whether or not to implement the proposal. It is also useful for summarising the intent of the proposal when it will be developed outside the workshop. The essential information includes the following:

    o   The proposal number

    o   Outline and more detailed description

    o   How it differs from the existing proposals (pre-workshop)

    o   Any change in functionality

    o   The advantages and disadvantages over existing proposals

    o   The likely impacts on cost, time and quality

    o   How it might be implemented

o    Any other comments or recommendations from the owner or those who developed it.

Finally, it contains a space for summarising the discussion at the wrap-up meeting and decision on whether or not to implement.

2.    The next five forms provide formats for the working papers that support the summary form. These are:

o    Sketch or description of the existing proposal

o    Sketch or description of the new proposal

o    Cost estimate of the existing proposal

o    Cost estimate of the new proposal

o    Whole-life cost comparison between existing and the new proposal.

3.    The final form is a more detailed analysis of how the proposal will be implemented, should it be accepted.

The study leader should allow a week or two after the workshop for the owners of ideas selected to work-up proposals, with others in the project team as necessary, to develop a properly validated and costed proposal. Proposals which had quality thought and development time going into their development are much more likely to be robust. If the decision is taken to implement them, they are less likely to unravel because of unforeseen details later in the project. This is a good reason to develop proposals after the workshop, regardless of the time available.

## 10.14 Scenarios

It is almost inevitable that some of the proposals will be wholly or partly mutually exclusive (in other words, one cannot implement both proposals because they have some common elements). Others will complement each other. A very effective way of selecting proposals to maximise the added value to the project is to group them into scenarios comprising a number of favoured proposals and to assess the impact they are likely to have on the value index calculated from the value profile. The method is as follows:

■    Select the best proposals for development by one of the methods indicated above.

- Group the proposals into scenarios being careful to identify any areas of mutual exclusivity.

- Assess the effect of the proposals on each of the value drivers.

- Compare the increase in the value index for each scenario.

- Review the affect of the scenario on the risk register.

By exploring a number of scenarios in this manner, it is possible to select, say, two, one which maximises value and the other which minimises risk.

## 10.15 Target costing

The price that customers are prepared to pay for something is dictated by the long-established law of supply and demand. If supply is scarce and demand is high the seller can dictate the price. If, however, competition is fierce between suppliers, only those with the most competitive offering will succeed. These conditions are typified by the automotive market in recent years. No longer can Henry Ford say 'they can have any colour so long as its black'. Manufacturers must provide the goods that the customer wants to buy at a price that *the customers* are prepared to pay. Value management plays a key role in establishing precisely what the customer wants and how much he is prepared to pay (see Value Management in Manufacturing in Chapter 7.4). Function cost analysis provides a key tool in establishing what functionality the customer wants. It also provides a means of assessing how much they are prepared to pay for the different functions. Using this base, the manufacturer can work with his designers and the supply chain to deliver the appropriate functionality at a price that the customer is willing to pay.

This approach is in marked contrast from that used where competition is not fierce and there is a seller's market. In this case, the seller adds up the total costs of supply, adds a margin for profit and establishes his selling price. In other words, the seller dictates the price.

$$\text{Cost} + \text{Profit} = \text{Price}$$

In a competitive environment the approach is different. Price is dictated by the customer. If the seller wants to remain profitable he must obtain the goods at a lesser cost, the difference being his margin for profit.

Price − Profit = Cost

This cost is the target cost. If the seller obtains the goods at a cost greater than the target cost, his profits will fall.

The construction industry has long been the victim of low margins and profits, resulting in many of the symptoms highlighted in the Latham Report. One of the reasons for this lies in the process of arriving at the price of the building. The architect designs a building, the quantity surveyor then prices it, based on historical rates, experience and market testing (all of which contain unstated allowances for risk). This approach is effectively the application of the first of the above formulae, Cost + Profit = Price. All too often the buyer is not prepared to pay the price and battle ensues to reduce it. Often this is achieved by 'negotiation' with suppliers (squeezing their margins), reducing the specification (reducing quality) and reducing scope (reduced functionality). The result can be a building which disappoints everyone.

Application of value and risk management from the outset of the project will establish the functionality required by the client (the buyer) and the necessary margins to sustain a healthy business. If the resulting cost, using the second formula, Price − Profit = Cost, is lower than the estimate, the function cost model will show where the mismatches are. The client can then work with the designers and suppliers to align cost with the required functionality without resorting to squeezing their margins. This results in a healthier and sustainable supply chain and an outcome in which expectations of all parties are satisfied.

The above approach will only work in an atmosphere of trust and collaborative working, where open book accounting is used. It is, therefore, closely allied with partnering.

## 10.16  Function performance specification

There are three types of specification which may be used to describe a designer's or a user's intentions to a provider of the item being specified. Each provides the supplier with differing degrees of freedom to offer innovative and competitive solutions.

Absolute specifications are those in which the specifier describes exactly what is required, in great detail (even down to specifying the maker of the item). This type leaves nothing to chance and gives the provider no freedom of choice. Scope for innovation is virtually nil. There is no opportunity for competition between providers except through their purchasing power or adjustment of margins.

Performance specifications describe the output required from a component or subsystem, but does not dictate the means of delivering that output. For example, the specification may require the internal environment in a room to be $21 \pm 2°C$ throughout the year with an occupancy of up to 50 people. There are a variety of air conditioning systems that could achieve this and the choice is left to the provider. This type of specification provides scope for innovation and genuine competitive bidding.

Function performance specifications (FPSs) take this one step further. The specifier defines user requirements by what the item must *do* rather than what it must *be*. The specifier (usually the user) expresses his needs in functional terms, without reference to possible solutions and with the minimum of constraints. In the previous example the specification might read 'the meeting space shall facilitate comfortable gatherings for the users and their guests at all times'. Here the provider has complete freedom of selection of his solution. The function is to enable meetings to take place – these could be in a room or they might be in some other area, such as an internal 'street'. The definition of 'comfortable' is flexible and could be negotiated with the users. Similarly, the number of people attending a gathering can vary. This gives the provider ultimate flexibility and ability to devise a very innovative and competitive solution.

The FPS comprises a thorough functional description of the system or subsystem being specified as it relates to the user (a user-related function, URF). To each function it attaches measurable attributes (known as Compliance criteria, CC). Each CC is assigned a target level (qualitative or quantitative). Each target level is provided a degree of flexibility by defining tolerance bands. Tolerance may be in four bands, zero, minimum, reasonable or plenty of flexibility. For example, the FPS for a paperclip might identify four primary URFs that it must perform: grip paper, not mark paper, have a stylish appearance and resist accidental detachment.

With each of these URFs will be one or more CC. For example, the function, grip paper might have a CC related to the number of sheets of A4 paper it will hold together. The level will describe how many sheets of paper it should grip, say 20. The tolerance band for the level may be 'reasonable', allowing compliance if the paper clip grips, say 20 sheets ±3.

The method is to undertake a function analysis of the users needs, identify measurable CC and agree the target levels and acceptable degrees of freedom. Having completed the FPS they can explore trade-offs between the CC for different functions. This enables them to arrive at the most competitive offering that complies with the specification.

# 11 Risk management techniques

## Summary

■ This chapter describes techniques that are primarily, but not exclusively, used in risk management.

■ First it describes the overall process for the management of risk in projects.

■ It then provides more detail of the key stages and gives details of techniques that may be used. The stages that are described in detail are identifying risks, assessing risk and managing risk.

■ The chapter then introduces some other techniques that may be useful in the management of risk. These include optimism bias, construction design and management (CDM), risk breakdown structures, failure modes and effects analysis, and hazard and operability (HAZOP) studies.

## 11.1 Risk management techniques

The following techniques are used mainly in the management of risk, although some may be used in other contexts.

## 11.2 Risk analysis and management

Chapter 3 outlined a seven-stage process for managing risk. This section provides a description of the stages.

### Stage 1 – preparation

Before attempting to manage risk, the study leader must have a full understanding of the project. The initial briefing process should therefore be very similar to that described earlier in Section 9.2.

There should be particular emphasis on understanding the environment within which the project is taking place since some of the greatest risks may arise from outside it. The external environment will include not just the built environment into which the building must fit but also strands of business activity (which may be competing for resources) as well as external stakeholders and authorities. During this stage, the study team leader should establish who needs to be consulted for the identification of risk (see below) and the most appropriate means of communication. Chapter 12 contains a checklist for issues to address during this stage.

## Stage 2 – identification

This is a very important stage and should not be rushed, since it is not possible to manage a risk unless it has been identified. Methods of identifying risks are discussed in more detail in the next section. Consultations should include those who represent interests from outside the project team. The process should identify the consequences of the risks should they occur.

A checklist for the identification of risks is included in Chapter 12; however, this should only be used towards the end of this stage to ensure nothing of significance has been omitted rather than at the outset, when it may be used as a substitute for original thought.

The study leader will enter the identified risks and their consequences onto a list, commonly termed the risk register. A suitable proforma for the risk register is included in Chapter 12.

## Stage 3 – analysis

Not all risks are of equal severity. It is therefore necessary to assess their relative impact on the project.

The first steps to group similar risks into appropriate categories (a suggested checklist of categories is included in Chapter 12 but other categories are in common use, for example, those given in HM Treasury's Orange Book).

The next step is to assess the likelihood of each risk, its consequence and the impact of the consequence upon the project. Methods of assessing risks are discussed in Section 11.4.

During this stage it is common for the study leader to agree with the project team the relationships between a qualitative description of likelihood and impact and the numerical probability and impact estimate represented by it. Impact estimates will be matched against ranges of cost (what effect will the consequence have on costs?) and time (what effect will the consequence have on activity or project durations?). Guidelines for these assessments are included in Chapter 12.

## Stage 4 – evaluation

The first step in evaluation is to agree whether or not risks should be quantified. Arguments for the justification of quantification are rehearsed in the section titled 'Quantifying risks'. It is not essential to quantify risks in order to manage them.

The relative severity of each risk may then be estimated by multiplying the likelihood and impact together to give a qualitative severity rating. Similarly, quantitative estimates of probability multiplied by cost or time estimates provide a quantitative estimate of the exposure.

The qualitative rating or the quantitative estimated exposure both provide a means of ranking the risks in terms of their severity. This ranking informs the team on their priorities for managing the risks, generally focusing most effort on the most severe risks. A further piece of data that may be available during the identification stage is the proximity of the risk (or when it might occur). If a risk is imminent it may be prudent to expend more management effort on this risk than on a more severe risk which is unlikely to occur for some time. Strategies for managing risks are discussed in Section 11.5.

If it has been agreed that risks should be quantified, it is at this stage that the study leader will calculate the risk allowances for cost and time. Methods for doing this are discussed in the section titled 'Quantifying risks'.

The study leader will enter the outcomes of the risk evaluation into the risk register.

## Stage 5 – treatment (or management)

The first step in this stage is to identify action owners for each risk (who will be responsible for managing them), what management action the owner will undertake to implement the agreed risk management strategy and a timetable for doing it.

The study leader will add the agreed management actions to the risk register. He will at this stage agree the risk management plan, describing how risks will be managed in the future, with the project team.

The study leader will compile a report describing the outcomes of these five stages as a record of the study and a recommendation for ongoing reviews.

## *Stage 6 – review*

At regular intervals, the study leader (or other person nominated in the risk management plan) should review the effectiveness of the risk management actions and update the risk register accordingly. Methods for undertaking reviews are given in Chapter 9.

## 11.3 Identifying risks

An extremely effective tool in building up a thorough understanding of the potential risks is to undertake a function analysis. This will identify the value drivers or primary functions for the project. The detailed processes are described in Section 10.2. Any shortcomings in delivering the value drivers will detract from the success of the project and therefore presents an area of risk. Risks may be directly linked to the function diagram.

It is common practice for the study leader to gather risks in advance of any workshop (see Section 11.2). The main advantage of gathering risks in advance is that the process is more thorough. The participants have time to mull over the potential risks and think them through, rather than be rushed into generating them during a workshop.

It also allows the study leader to group and assess the key areas of risk so that he can direct the workshop accordingly (see below). Gathering risks in advance allows the team to use valuable workshop time to agree management strategies and actions rather than simply generate a list of risks.

The risk identification strategy will be agreed during the briefing process and should be consistent with the guidelines on the level of activity that is justifiable for the project (see Section 4.13).

When identifying risks the study leader should record not just the risk (or threat) to the project (described adequately so that third party can understand it) but also the consequences (or effect) that would follow should the risk occur. It is the consequence that will have an impact on the project, not the threat of the risk occurring.

## Interviews

The most effective way to gather risks is to speak to the stakeholders. Interviews are more fruitful if the study leader sends to all the interviewees an outline description of the proposed risk study and a form showing the proposed categories under which risks will be grouped, for example, those shown in Table 6.6. The guidelines will help the interviewee gather his thoughts in advance of the interview. Interviews may be one-to-one or with small groups of people with similar interests in the project. It is usual to assess the severity of the risks during the interviews (see below). The drawback of the interview technique is that it can take up a lot of time. Its use therefore may not be justifiable on smaller projects.

## Questionnaires

The next most thorough way to identify risks in advance of any workshop is to send the stakeholders a questionnaire, setting out the same information as you would use to structure your interviews, together with a description of why they are being asked to complete it. The questionnaire should include the proposed risk categories and guidelines on assessing the severity of the risks. Shortly after sending out the questionnaire the study leader may telephone the recipients to answer any queries and help them complete the questionnaire while on-line. This emulates the interview technique but takes up significantly less time.

## Workshop

If neither of the above techniques is practical the study leader is left with the commonly used technique of generating risks during a workshop (either specially convened for identification and assessment of risks or as part of a broader workshop which addresses risk management actions and other topics). The traditional technique is to brainstorm the risks to generate a comprehensive list. Once the list is complete the workshop team will assess them separately.

An alternative to this technique is to prepare a large (A0 size) risk evaluation matrix, reflecting the impact and likelihood levels to be used on the project and fix it to the wall. The study leader then asks the workshop team to write the risks they identify on self adhesive notes and place them in the appropriate square in the matrix. This is quick, interactive and can be fun. It works best with relatively small workshops of 12 or less people.

All the methods of identifying risk will be informed by function analysis. Giving the team a simple model that describes the project imperatives (for this is what Function Analysis does) helps them to think of what might go wrong. Indeed one way to structure the identification of risk is to use the function model.

There are numerous techniques which may be found in the textbooks for identifying and assessing risks. Most of these are seldom used on building projects. Some of the more useful techniques are described below. These techniques may be used in conjunction with any of the other identification techniques described above.

## Prompt lists

The first, and probably most commonly used, is the 'prompt' list. This list is built up, from experience and contains all the likely risks that may occur, grouped under appropriate categories. Many organisations have developed their own preferred lists to suit their way of working. A prompt list categorised under the same headings as those given in Section 12.8 provides groups of things that could cause project failures.

Prompt lists should be used with caution. They should not be used as an alternative to original thought. They should be used at the *end* of the risk identification process as a safeguard that the process has been thorough and explored all likely areas of risk.

They should be used only to generate risks that are *specific* and *relevant* to the project and to the stage in the project lifecycle at which the study is being conducted. For example, on many projects there is a risk that the contractor may encounter unexpected obstructions in the ground. The study leader should ask the question: 'Is there an abnormally high risk of this happening on *your* project?' If not, the risk may not be worth recording, particularly in the early stages. The danger of recording every risk is that of generating overly long lists of generic and trivial risks that do not require (and resource limitations would never permit) active management. Such lists should be avoided as they undermine effective risk management by hiding the severe risks that *do* require attention in a fog of trivia. They are also a huge turn off for those involved. Risk reviews then become dreaded, tedious and low priority events and soon cease altogether.

## Cause and effect diagrams

Cause and effect diagrams are a useful analytical way of linking and identifying risks and may be used by individuals or small groups in preparing their response to contribute risks. Their build up is not dissimilar from the logic of a function tree. Starting with a consequence, the respondent asks the question 'what might cause this consequence? (or *Why* would it occur?)'. He then looks at the immediate cause and repeats the question. The answers will quickly build into a herringbone diagram as illustrated in Figure 11.1.

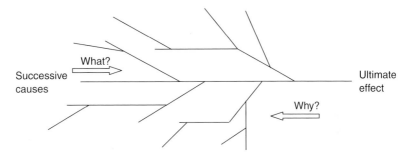

**Figure 11.1** Cause and effect diagram

Alternatively, the respondent might start with a perceived risk and explore the consequences of that risk occurring by asking the question '*What* could happen if ... occurred?' Successive repetitions of this question will result in the identification of consequences of increasing severity.

The diagram is likely to have several branches relating to different aspects of the project. These could be aligned with the preferred categories under which risks will be grouped.

Because the effects (towards the right-hand side of the diagram) increase in severity the cause and effect diagram can be a useful way to filter out the most severe risks to the project and avoid the long lists of risks that can so easily be generated.

## Decision tree

Another useful technique is the decision tree. This is similar in structure to the cause–effect diagram except that it links the decisions that might lead to a particular outcome. As indicated in Figure 11.2 the decision tree allows the respondent to assign the probabilities of each division being made (such that the total probability for each choice always adds to 1). This can then be used to ascertain overall probabilities of risks occurring.

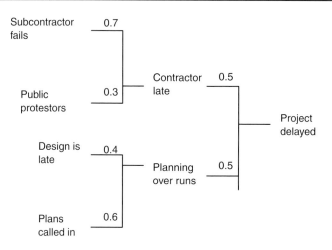

**Figure 11.2** Decision tree

## *Addressing strategic risks*

In the early stages of a project, the management prefers to focus on the main strategic risks to help them decide two things: whether or not to commit to a project and; if so, what project do they want to commission. A very effective method to focus on strategic risks at this stage is discussed in the section titled 'conventional workshop' in Chapter 8 as part of the project definition study. Typical questions to probe whether clarity in the definition exists are set out in Table 11.1.

**Table 11.1** Typical questions for creating the conditions for success

| Risk category | Typical questions |
| --- | --- |
| 1.0 Business and funding issues | Is funding in place? Has a written business case been prepared? Are the resources in place for the project to proceed? |
| 2.0 Clarity of brief | Does everyone understand exactly what the client expects to build? Have you prepared a current written current brief, is it clear and do you understand it? |
| 3.0 Management roles and responsibilities | Does everyone understand their own role and responsibilities? Is the client body clear? Are all appointments in place? Who is responsible for coordinating between the different disciplines/packages? Is all information/drawings consistent? |

## 11.4 Assessing risk

The basis of assessing risk is to explore how likely they are to happen, what the consequences of their occurrence will be and the magnitude of their impact in terms of time, cost or quality.

## *Qualitative assessment*

For each risk that is identified the team need to assess its severity in order to decide what course of action to take. They assess severity by looking at the likelihood that the risk will occur and the magnitude of its potential impact. This can be done qualitatively or quantitatively. The qualitative severity rating is arrived at by multiplying the qualitative impact by the likelihood of occurrence.

## Use of matrices

The use of $3 \times 3$ or $5 \times 5$ matrices for assessing the severity of risks was discussed in Section 3.9. Both suffer from the disadvantage of a middle point in each of the ranges. This encourages a 'cop out' during assessment by picking the middle of the range. Better would be a 4-point scale which forces the assessor to choose an impact or a likelihood on either side of the midpoint, although this is seldom used.

The $3 \times 3$ and $5 \times 5$ matrices also suffer from another disadvantage. There is no distinction between the risk rating for a risk which is highly likely but low in impact or extremely unlikely but high in impact. In the above examples such risks would each score 3 or 5, respectively.

Thus a risk whose impact could be disastrous but is extremely unlikely (it could collapse) is rated the same as one which is of negligible impact but highly likely (it will probably rain). Those managing risks should be paying more attention to the former than the latter.

*Note*: Although the above assessment uses units to arrive at a numerical severity rating, these numbers bear no relation to the physical severity of the risk in terms of cost or time. They are simply a non-dimensional score to assist in ranking risks based on qualitative measures of impact and likelihood. This distinguishes them from quantitative assessments of severity discussed in the next section.

## Skewed matrices

The problem can be overcome by using a skewed matrix, which rates impact higher than likelihood. One such matrix is based on the Risk and Management of Projects (RAMP) system devised by the Institution of Civil Engineers and Institute of Actuaries.

In this matrix, likelihood is measured on a scale of 1–16, whereas impact is rated on a scale of 1–1000. Thus a risk which is highly likely but of marginal impact will rate 16. However, if it is extremely unlikely but could have disastrous consequences, it will score 1000. The RAMP system recommends that any risk scoring 1000 or more must be actively managed.

It should be noted that in all of the above systems the rating numbers are dimensionless and simply represent scales for establishing the severity of a risk. They are not related directly or indirectly to units of time or cost.

The qualitative matrix may also be used as a guide to estimate the cost and time impacts resulting from a risk occurring. Tables 11.2 and 11.3 relate the qualitative descriptions for likelihood to suggested ranges for cost or time probabilities. Similarly, impacts are expressed as percentages of the total project cost and total project duration respectively.

It must be emphasised that these ranges are only examples. The actual ranges used must be agreed with the project team to suit the project risk profile and the organisation's appetite for risk.

**Table 11.2** Likelihood assessment table (based on the principles of the RAMP system), published by Thomas Telford

| Description | Scenario | Code letter | Score | Guide probability (%) |
|---|---|---|---|---|
| Highly likely | Very frequent occurrence, almost certain | HL | 16 | Over 95 |
| Likely | More than even chances | L | 14 | 50–95 |
| Fairly likely | Quite often occurs | FL | 10 | 21–49 |
| Unlikely | Small likelihood but could well happen | U | 6 | 2–20 |
| Very unlikely | Not expected to happen | VU | 3 | 0.5–2 |
| Extremely unlikely | Just possible but very surprising | EU | 1 | Up to 0.5 |

**Table 11.3** Impact assessment table (based on the principles of the RAMP system), published by Thomas Telford

| Description | Scenario | Code letter | Score | Guide cost % of project | Guide time % of programme |
|---|---|---|---|---|---|
| Disastrous | Business investment could not be sustained, project at risk | D | 1000 | >2 | >5 |
| Severe | Serious threat to project | S | 550 | 1–2 | 2–5 |
| Large | Reduces viability significantly | L | 60 | 0.5–1 | 1–2 |
| Marginal | Small effect on viability | M | 11 | 0.1–0.5 | 0.5–1 |
| Negligible | Trivial effect on viability | N | 1 | <0.1 | <0.5 |

## Workshop assessment

It is not always practical to identify and assess risks by means of consultation. In these circumstances it may be necessary to convene a risk identification and assessment workshop. This may not be as thorough as the consultation process but it can be made vastly more interesting and effective by using the following technique. After the workshop preambles, the study leader affixes an A0 size copy of the matrix shown in Figure 11.3 (often referred to as a heat map because of the colours of the squares), to the wall and asks each member of the team to write (in bold marker pen) their top risks on separate 'Post-it' notes. He then asks them to explain the risks and stick the notes in the appropriate square on the matrix. He allows the group to discuss and move/add new risks as necessary.

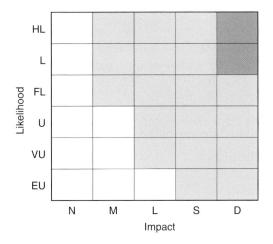

**Figure 11.3** Qualitative assessment matrix or heat map (For description of letters used in the figure see Table 11.3)

This method enables identification and assessment quickly and builds the team's interest in the process. The risks and their assessments can then be entered onto the draft risk register.

After a short time, the matrix will contain a very comprehensive array of risks, together with their assessments (which have, of course, been agreed by the team). The workshop recorder can then enter the risks into a register. Later, the study leader can sort the risks into agreed categories. He can address the most severe risks (those in the top right-hand corner of the matrix) during the workshop and ask the team to agree the appropriate management actions against them.

## Quantifying risks

In some circumstances it is desirable or necessary to quantify the risks to a project in terms of cost and/or time. Reasons for quantification could include:

- Where clients need to report upwards in their organisation or to a third party.

- Where the project forms part of a larger programme of projects.

- To focus people's attention so that they follow through management actions.

- Where clients insist on it as part of their procedures or have capped funds.

- Where it is desirable to link risk to contingency.

- Where it is required or provides comfort to funding organisations or other third parties.

Quantification of risks will result in an estimate of an allowance that should be added to the base estimate of cost and/or time and will have the dimensions of cost or time.

### Statistical analysis

Simple multiplication of the impact (expressed in cost or time units) by the probability of occurrence (expressed as a percentage) results in the 'nominal exposure' for that risk. Adding the nominal exposures of all risks within the risk register across the project gives a figure for the risk allowance. However, statistically, there is only a 50% chance that this risk allowance will be sufficient (or, in other words, there is a 50% chance that the project cost will overrun this risk allowance). Calculation of a risk allowance giving a higher level of confidence requires a more sophisticated and statistically sound calculation method.

## Central Limit Theorem

The Central Limit Theorem provides a simple spreadsheet-based method of calculating risk allowance with a 90% confidence level that will be sufficient. The formula upon which this is based was developed by the mathematician Laplace in the nineteenth century.

For a risk allowance giving 90% confidence, the formula is:

$$\sum P_i \times E_i + 1.3 \times \sqrt{\sum (E_i^2 \times P_i) \times (1 - P_i)}$$

where $E_i$ = the estimate for risk $i$ and $P_i$ = the probability of risk $i$ occurring.

The formula calculates the risk allowance that should be added to the base cost estimate to give 90% confidence that the project can be completed within the resulting sum.

To apply this method the study leader should:

■ Agree the ranges for converting the qualitative assessment of likelihood and impact with the team during the workshop.

■ Outside the workshop enter the probabilities and estimates against *all* risks included on the quantitative register (not just those with management actions).

■ Calculate the nominal exposures and the factor to provide 90% certainty.

The risk allowance calculated by this method provides an estimate of the sum that is needed to provide 90% confidence in delivery, *assuming that the risks are not mitigated*. It may well exceed normal expectations of contingency at that stage in the life of a project because some management actions are not complete. It will focus attention on the management effort needed to reduce the risk allowance.

The Central Limit formula assumes near normal distribution of risks and is based on a 'risk free' estimate. When using the method the study leader should ensure that the base cost estimate is risk free, or make allowances for any risk that has been build into the rates, otherwise there will be double-counting and the estimate will be higher than necessary.

## Computer-based simulation techniques

In order to achieve a range of risk allowances linked to different confidence levels it is necessary to analyse the risks using computer-based software running Monte Carlo simulation or similar programmes.

The procedures for identifying risks for subsequent analysis by the Monte Carlo technique are exactly the same as described above except that it will be necessary to statistically model the chances of each risk occurring and its impact should it occur. The initial risk register can be used as a basis to undertake a Monte Carlo analysis.

The following additional information is necessary to run a Monte Carlo analysis:

- The probabilities of each risk occurring.

- Information to statistically model the impact of each risk should it occur. For example, the minimum, most likely and maximum costs to create a triangular distribution. The most commonly used distributions include:

    o   Uniform – There is an equal chance that the parameter will have any value between two limits (e.g. the cost per tonne will be between £X and £Y).

    o   Triangular – The minimum value of the parameter is X, the most likely is Y, the maximum is Z.

    o   Discrete – Either the event happens or it does not.

An operator, trained in the use of an appropriate computer software program should then enter the data and run the program.

Computer-based analysis software can generate the following outputs:

- Probabilities of project completion at various costs. For example, 90% confidence that the project can be completed for less than £xxx. For a uniform spread of risks, this is consistent with results given by the Central Limit Theorem described above. (The study leader is not limited to the 90% confidence figure. He can select the confidence level to suit the circumstances.)

- Distribution of outturn cost outcomes, for example, most likely cost outcome or the expected outturn cost.

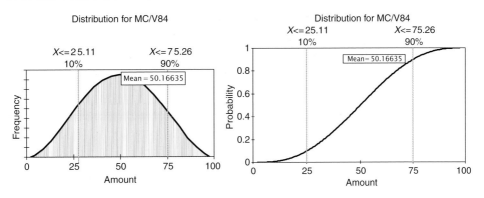

**Figure 11.4** Computer generated cost and probability histogram

■ Identification of the risks that have the most impact on the project outcome.

Other statistical information can also be generated.

*Warning!* The results of this analysis will only be as good as the information upon which it is based. Since most uncertainty estimates will be subjective, the accuracy of the results will only be approximate. This is commonly known as the garbage in, gospel out (**GIGO**) syndrome.

Most computer-based risk programmes allow the use of a large variety of probability distributions to describe each risk. The quality of the input data is generally not sufficiently reliable to warrant the use of many of the more sophisticated ones.

## Time risks

Because activities within a project may or may not lie on the critical path (see Section 3.10), it is not possible simply to add uncertainties of duration to obtain a time-risk allowance. It is therefore essential to run a computer-based simulation programme. This requires that, in addition to identifying each time risk, it is necessary to identify which activity in the critical path programme it might affect.

The study leader should enter the time risks into the software program (usually a type of spreadsheet). The software will then run a Monte Carlo simulation to analyse the time risks. The software can generate the following reports (among others):

■ Graphs showing the total project duration against probability of achieving it.

■ End dates against selected probabilities of achieving them.

■ Identification of the risks that are the most severe and therefore need most management attention.

*Warning!* Because this requires interaction with a logic-linked critical path programme, the study leader should check the quality and ownership of such a programme before proceeding.

## 11.5 Managing risk

### *Guidelines for selecting risks for active management*

Once the team has agreed on the initial risk register and the risk management strategies, the study leader should agree with the team on which risks require active management. It is not necessary to quantify risks in order to put in place an active risk management programme.

### *Selecting the range*

It is neither practical nor necessary to actively manage all risks on the risk register. Only those which represent the greatest threat to the project require this treatment. To select risks for active management to the team agrees a threshold severity rating above which the team will actively manage risks. This avoids unnecessary effort in managing trivial risks and focuses the effort on the most significant risks. The threshold will vary from project to project and will depend on the team's risk appetite. By using the scales described earlier in this section, risks with a rating above 500 or 1000 should be actively managed. Alternatively, Table 11.4 provides a guideline for selecting risks for active management.

**Table 11.4** Rating recommendations (based on the scales described in the section titled 'Qualitative assessments')

| Rating | Description | Action required | Strategy | |
|---|---|---|---|---|
| Over 5000 | Intolerable | Must eliminate or transfer risk; report to management board | Eliminate | Terminate |
| 1000–5000 | Highly undesirable | Attempt to manage, avoid or transfer risk; report to project board | Reduce | Treat |
| 100–1000 | Manageable | Retain and manage risk; record status of project progress reports | Insure | Transfer |
| Up to 100 | Negligible | Need to actively manage; review from time to time | Contain | Tolerate |

## Escalation

Table 11.4 above also provides guidance on the levels to which the status of the risks should be reported. Those risks that could threaten the future of the project or, indeed, the business must be reported to the highest levels of management. Risks of lesser severity should be reported to a level of management consistent with the level of threat. This mechanism for filtering out only the most severe risks for escalation ensures that senior management reports are not overburdened with trivial risks that could distract them from their executive responsibilities.

Once one or more management action(s) have been agreed, the study leader should record the action and a named individual as the action owner. A named individual is better than an organisation, since it makes the individual more accountable for leading the management action.

The study leader should record a date by which the action should be completed or reviewed. He should not allow the word 'ongoing' to appear in the date column of the register. The word 'ongoing' invites prevarication. It is better to identify the next review date, say the next project team meeting, at which the action owner can report progress.

## Risk management strategies

There are four major strategies for dealing with risk, and these are referred to in literature under various acronyms. Some use the acronym ERIC, described below. A commonly used alternative acronym is TTTT. This is shown in the parenthesis below.

E*liminate (Terminate) show-stoppers and biggest risks.* This action will be taken only with the most severe risks which threaten to cause project cancellation or failure. For these risks it is essential that the risk be removed or converted to enable one of the other strategies to apply. There are several ways to do this: the team could review the project objectives and remove the objective which contains the intolerable risk. The result will change the project. They could reappraise the whole concept of the project, again creating a changed project. They might decide that the risk is so great that it is decided that the project is not a viable proposition. This would result in cancellation. Summarising, a strategy to eliminate or terminate a risk will result in significant change to, or cancellation of, the project.

*Reduce (Treat) risks.* This is achieved by undertaking surveys, redesign, use of other materials, use of different methods or by changes in the procurement plan. The approach is less drastic and more common than elimination. Reduction is accomplished by actions undertaken by members of the project delivery team, for example, the design team can change the design to eliminate that bit of the design which causes the risk. Such change might result in increased cost. The team can undertake surveys to provide better information thus removing uncertainty and thereby reducing the risk. Surveys will cost money to conduct. The team can review different materials or equipment to perform the required outcome. By adopting a more proven method or solution (which by definition is less innovative) the risk is likely to be reduced. They could propose different ways of working towards the project objectives and reduce risk in that way. They could package the work in different ways to reduce the interfaces between the trade contractors thereby reducing the risk. The effect of reducing or treating risks is generally to increase the base cost and reduce the risk allowance because the team is paying to undertake the risk reduction process.

*Insure or (Transfer) risks.* Some risks can be insured or transferred to other parties. Insurance is a straightforward transaction where by paying a premium, should the risk occur, the insurance company will recompense the damage caused. This is another commonly used strategy, generally achieved by allocating risk to the contractor through the contract. Allocating the risk in this manner does not reduce the risk. It normally results in the client paying a premium to transfer risk to the contracted party. Adopting this strategy results in paying a premium to reduce the risk allowance.

*Contain (Tolerate) the risk within the unallocated contingency.* This is the strategy to adopt for all minor risks. It involves no active management. The team judges that, should the risk occur, the cost is affordable within the risk allowance or contingency fund allowed for the project. Because these risks are less severe the competent management team will feel comfortable about addressing them only if they occur. Only then will they have an impact. However, if they occur, the full impact of the risk must be borne by the party to whom that category of risk is allocated through the contract.

There is a common misconception that transferring a risk to another party in the project delivery supply chain eliminates the risk. This is, unfortunately, not the case. The contractor, or other party to whom the risk is transferred, must still manage the risk. The risk is still present. Should they fail to eliminate it, they must pay the consequences. If they cannot, the impact of the risk reverts to the client body.

## Allocating management actions

Which strategy to adopt for the risks that have been identified will depend upon their severity (the product of likelihood and impact) and the team's risk appetite. For each risk it is necessary to consider who is accountable should that risk occur. This person is normally called the account owner, and will be a senior manager or board member. The team must also decide on who can best manage the risk, either on his own or in collaboration with others. This person is normally called the action owner. Next the team needs to consider what the action owner can undertake to implement one of the strategies outlined above. This will be the management action. Finally, the team needs to decide by when the action should be completed and when it should be reviewed.

In defining the action that the action owner should take, it is necessary to keep things in proportion, assess the resources needed to undertake the action and compare these with the impact should the risk occur. There is little point in expending more resource to manage a risk than to its impact, should it occur.

## Updating the risk register

All the above information should be entered in the risk register, a suitable format for which is shown in Section 11.9. (Note that in this figure the term owner refers to the action owner described above, not the account owner.)

The risk register forms the basis of the risk management plan. It should be reviewed at regular intervals by the risk manager and updated to reflect the effectiveness of the management actions and (hopefully) the reduction of risk. It should be reported through the regular project management reporting channels.

## To quantify or not to quantify

It is not necessary to quantify risks in order to manage them effectively. Quantification, particularly on construction projects, is notoriously subjective and inaccurate. This is largely due to the virtual total absence of any actuarial information on those events which can cause the greatest impact. This, in turn, is largely due to the individual and non-repetitive nature of construction projects.

The risk manager should only quantify risks if there is a clear benefit from doing so. Quantification of impact is normally expressed in terms of cost or time or both. Quantification of likelihood is expressed as a probability. Neither of these need necessarily be single point estimates. Quantification can be expressed as a range of values, and likelihood may be expressed as a probability distribution.

In order to understand the overall impact on a project, it is necessary to address all the risks within the risk register and use an appropriate statistically sound method to assess the overall impact of risk on the project. Such quantification is explained below.

The quantified risk register must be reviewed on a regular basis, adjusted as necessary to reflect the effectiveness of the actions being taken. The allowances made for total project risk must be regularly reviewed in terms of cost and time.

If the quantified risk register is linked to the project management contingency, quantification of risks provides a very powerful project management tool for controlling the project. Only if an identified risk occurs should contingency be released. If an identified risk does not occur contingency is not released. This concentrates hearts and minds on two things, first the thorough identification of potential risks and second on ensuring that they do not occur. This is where the quantification of risk can be very useful.

Where a number of projects are being undertaken simultaneously, quantified risk allowances enable the contingency in one project to be transferred to other projects within the programme, once the risks have been managed or have passed. In a large portfolio of projects this enables more projects to be undertaken. Once part of the risk allowance in one project has been transferred to another, the team must remember that it is no longer available to the project from which it was transferred.

## The risk register

The risk register is a key project management document and describes the status of the risk management process.

Having identified the risks, the study leader will enter them on an initial risk register, grouped under the appropriate category, amalgamating duplicated risks so as to avoid repetition and double counting. The risk register will show:

- The identity of each risk and its unique identification number

- The category under which it is grouped

- Its consequence, should it occur

- Its severity assessment in terms of likelihood and impact and consequent rating

- The risk action owner

- The actions he will take and the date by which the action should be completed or reassessed

- The basis for quantification of risks, if appropriate

- The probability, impact estimate and risk allowances if risks are quantified.

The word 'on-going' for the action by date should be avoided. Instead, identify when the next date is on which this issue can be reviewed.

Sometimes the register will show the team's estimates of the severity of each risk after the management action has been taken. This judgement assumes that the action is successful and should be treated with caution. An extract from a typical risk register is shown in Table 11.5.

## Risk allocation

In the early stages of a project, before contracts are let, the team can use either a generic prompt list or a project specific risk register to allocate risks between the key project stakeholders. This document can then be used in the drafting of contract documentation to allocate the risks in the desired manner.

Several forms of contract allow for the allocation of risk to be specified in this manner, for example, the New Engineering Contract (NEC), published by the Institution of Civil Engineers and PPC2000, published by ACA Publications. This represents a much more equitable way of dealing with risk and, by defining the allocation, reduces uncertainty for all parties.

The risk allocation document is a key component of the tender documentation for PFI/PPP projects. One of the key principles

**Table 11.5** Extract from a typical completed risk register

| ID No. | Description | Consequence | Owner | Management action | Date by | Like-lihood | Impact | Rating | Comment | Ultimate owner |
|---|---|---|---|---|---|---|---|---|---|---|
| 1.00 | *Clarity and understanding of scope* | | | | | | | | | |
| 1.01 | Scope of facilities not finalised | Cannot complete design brief or reconcile costs with budget | AB | Develop building brief (identify stakeholders, process, aspirations, functions and potential solutions) Develop operational brief | 01-Feb-02 | FL | D | 9500 | | SPV |
| 1.02 | Client undecided as whether to include certain facilities | Will affect site layouts, suitability and costs | MF | Meet with users | 05-Mar-02 | FL | S | 5225 | | SPV |
| 1.03 | Sizes of centres not fixed | Will affect site layouts, suitability and costs | See 1.01 | | | FL | L | 570 | | |
| 2.00 | *Management structure and communications* | | | | | | | | | |
| 2.01 | Failure/delay in passing legislation | Cannot sign contracts so project delayed | MC | Address at steering meeting | 24-Jan-02 | FL | D | 9500 | | Client |
| 2.02 | No overall clientside project owner | Lack of leadership and decision-making | PJ | Meet with directors to agree selection | 30-Jan-02 | L | S | 7425 | | Client |
| 2.03 | Overall project governance plan not finalised | Confusion and misunderstandings | MC | Develop plan | 15-Mar-02 | L | 5 | 7425 | | Client |

of PFI/PPP is for the public sector to retain only those risks that pertain to its core business of service delivery. All risks relating to the design, construction and management of a building, from which the services are delivered, are allocated to the private sector contractor. Table 8.3 illustrates an extract from a standard matrix for the allocation of risk in PFI contracts.

## The risk management plan

The resulting management actions are entered into the risk register and form the core of the risk management plan. The account owner is accountable for ensuring the implementation of the plan. He is likely to be the project director or someone of similar executive standing. He may appoint a small representative group, a steering group, to regularly review its status and agree changes. He will also appoint a risk manager, responsible for the day to day monitoring of the management actions, convening regular review meetings and compiling reports. This person could be the study leader. The individual management actions will be undertaken by the action owners agreed earlier.

The risk manager will agree with the project director and the steering group on the overall timetable, dates for reviewing and reporting progress, the mechanism for reporting progress and escalating risks, and dates or milestones for formal risk reviews.

## Agreeing the plan

Once proposals have been developed and management actions proposed it is necessary that management make an informed decision as to whether to implement the proposals or not. The best forum for doing this is a specially convened meeting, called a wrap-up meeting, at which the owners of the proposals will present their findings to a management panel, described earlier in this chapter.

The plan must have an owner, generally called the risk manager. This is usually an existing member of the project team, with the authority to manage the plans, or it could be the study leader who conducted the original study. The latter has the advantage of being objective in managing the follow-up action, but suffers the disadvantage of not being in day-to-day contact with the project. The risk manager may form a small group of people from within the project team to assist in regular progress reviews. The outcome of each of these progress reviews should be summarised in a report to be included in the regular project reports and circulated to management.

If the risk manager feels that the management is not taking the follow-up seriously, he should make his views known to the accountable officer and raise the profile of the activity. The plan should have a clear mechanism for escalating risks to a higher management level if the outcomes of the planned actions warrant such action.

## Implementing the plan

It is a common failing among project teams to regard value and risk management studies as one-off events which, when passed, require only intuitive follow-up. This risks losing much of the value created during the study. Those responsible for implementing the plan should have full senior management support and the necessary resources to carry out their tasks.

Implementation activities resulting from value and risk management studies will depend upon the project stage at which the study is undertaken.

## Inception

At inception stage the main output will be to develop supporting information to an outline business case. This will assist senior management in making the decision on whether or not to proceed with the project and, if so, what form it might take. It is at this stage that they will decide whether the benefits they are seeking can be best delivered through a building solution or by some other means. The supporting information will confirm, or otherwise, that the project fits with the organisation's strategic objectives and can satisfy the requirements of the key stakeholders. It will summarise the key benefits that the project should generate and suggest a number of alternative ways in which such benefits may be delivered. Each of these options will have advantages and disadvantages. These may be expressed comparing their performance against a strategic functional performance model (in essence a business value driver model). The output will include an assessment of the risks both to the business and the potential project and suggest a plan for managing them.

## Concept

Once the senior management has decided to proceed with a building solution, they will appoint their professional team to begin developing the concept. It is at this stage that they might choose to undertake a project launch study to assist in defining the project requirements and to identify the key threats to the project. The main outputs from such a study will be a detailed value drivers model, linking the project objectives to design consideration, and a high-level project risk register, focusing mainly on project management issues rather than construction risks. Follow-up activities will therefore comprise embedding the messages from the value driver model into the project brief and ensuring that the project execution plan includes the necessary procedures to address the key areas of risk identified in a risk register. It should also include a summary of the on-going risk and value management plans for the remainder of the project and how progress will be reported.

## *Design*

As the design development proceeds, the team will develop a small number of options to satisfy the agreed concept. Value and risk management can assist in the selection of the most appropriate options by assessing their fit with the agreed value drivers and the levels of risk associated with each. At this stage follow-up will comprise validating and maybe improving upon the selected option (by conducting subsequent value engineering studies) and embedding the outputs into the design brief. The team will also undertake the actions identified on the risk register and review will progress at regular intervals.

Once designs have been developed to elemental level (say Stage C or D on the RIBA plan of work) the team should conduct a value engineering study. Follow-up from this study will be the full validation and development of the accepted proposals for improvement and their inclusion in the developing design. Regular reports should include the value-added cost and time improvements for each proposal. Risk management follow-up will involve a continuation of the risk reviews outlined in the previous paragraph.

The team may undertake a design and cost review. Follow-up from this review will be to incorporate the agreed proposals into the designs and specifications to redress the balance.

## 11.6 Optimism bias

At the early stages of a project, there is often insufficient time or information available to undertake a rigorous analysis of risk. Those proposing a project are habitually overoptimistic in their

**Table 11.6** Recommended adjustment ranges for optimism bias

| Project type | Works duration Upper (%) | Lower (%) | Capital expenditure Upper(%) | Lower (%) |
|---|---|---|---|---|
| Standard buildings | 4 | 1 | 24 | 2 |
| Non-standard buildings | 29 | 2 | 51 | 4 |
| Equipment/development | 54 | 10 | 200 | 10 |
| Outsourcing | N/a | N/a | 41 | 0 |

assumptions, leading to underestimation of costs and time and overestimation of benefits. Recognising this, HM Treasury commissioned Mott MacDonald to undertake a survey of the size and causes of overruns on past projects due to this 'optimism bias'. The results form the basis of a top-down process for estimating a risk allowance, without the need for undertaking a full risk analysis. The Treasury Green Book sets out guidelines for those preparing business cases so that they may add suitable amounts to the initial estimates to take account of this overly optimistic tendency.

The Green Book identifies upward adjustments for various construction types to which this applies as indicated in Table 11.6. The upper amounts are applicable at the outline business case stage. The lower amounts are applicable at the final business case stage, assuming that there have been improvements in the accuracy of costing and the management of risk.

If there has been a degree of formal risk analysis and management, the optimism bias adjustment should be reduced by an appropriate amount to avoid 'double counting'. This may be assessed using the following method. Table 11.7 gives the breakdown by contributory factor of the recommended adjustments in Table 11.6 for building projects assuming that they are unmitigated. Each of these may then be assessed against the circumstances prevailing for the project under consideration and the degree to which the appropriate risks have been set off or managed. For each contributory factor a mitigation factor between 0 (no mitigation) and 1 (completely mitigated) is assessed. The product of the mitigation factor and the unmanaged contribution from Table 11.6 gives the amount by which the optimism bias contribution for that contributory factor may be reduced.

Thus, taking the cost adjustment for inadequacy of business case for a standard building as an example and assuming that the

**Table 11.7** Optimism bias upper bound guidance for building projects

| Upper level of adjustment | Contributory factors | Non standard buildings | | Standard buildings | |
|---|---|---|---|---|---|
| | | 39% Works duration | 51% Capital expenditure | 4% Works duration | 24% Capital expenditure |
| Procurement | Complexity of contract structure | | | | |
| | Late contractor involvement in design | | | | |
| | Poor contractor capabilities | | | | |
| | Dispute and claims occurred | | | | |
| | Information management | | | | |
| | Other | | | | |
| Project specific | Design complexity | | | | |
| | Degree of innovation | | | | |
| | Environmental impact | | | | |
| | Other | | | | |
| Client specific | Inadequacy of business case | | | | |
| | Large number of stakeholders | | | | |
| | Funding availability | | | | |
| | Project management team | | | | |
| | Poor project intelligence | | | | |
| | Other | | | | |
| Environment | Public relations | | | | |
| | Site characteristics | | | | |
| | Permits/consents/approvals | | | | |
| | Other | | | | |
| External influences | Political | | | | |
| | Economic | | | | |
| | Legislation/regulations | | | | |
| | Technology | | | | |
| | Other | | | | |

mitigation factor has been assessed as 0.4, the revised optimism bias adjustment would be:

$$31\% - (31\% \times 0.4) = 18.6\%$$

If the estimated capital cost were £100 million, the contribution towards optimism bias of this one contributory factor, inadequacy of business case, would be £1 8600 000.

If a formal risk analysis has been undertaken and a risk allowance calculated, using a recognised statistical method, the optimism bias adjustment should be reduced. The Ministry of Defence recommend that the optimism bias adjustment should be similar in magnitude to the uplift from 50% confidence levels to 90%.

## 11.7 Construction, design and management regulations

Construction, design and management (CDM) regulations were introduced in 1994 to reduce the incidence of accidents and occupational ill-health arising from construction. To implement the regulations, which apply to all significant construction projects, a planning supervisor is appointed. His role is to coordinate health and safety issues throughout the design and planning stages. This role passes to the contractor once construction begins. The regulations not only require that the risk of accidents and ill-health are minimised during construction but also when the building is in use.

Risk management discussed in this book relates mainly to that of improving the likelihood of a successful project outcome. A part of that success is that the building is constructed safely in an accident-free environment and, after commissioning and handover, is a safe and healthy place to occupy. The planning supervisor, therefore, has a direct interest and a clear role to play in the risk management process.

The study leader should invite the planning supervisor to contribute to the risk management process to ensure coordination of effort.

## 11.8 Risk breakdown structures

The normal format for a risk register is a table, or matrix, listing the identified risks within specified categories, agreed by the project team as being suitable for their project.

Another way to present risks is to build a risk breakdown structure (RBS). This allocates each of the risks to the same categories as a

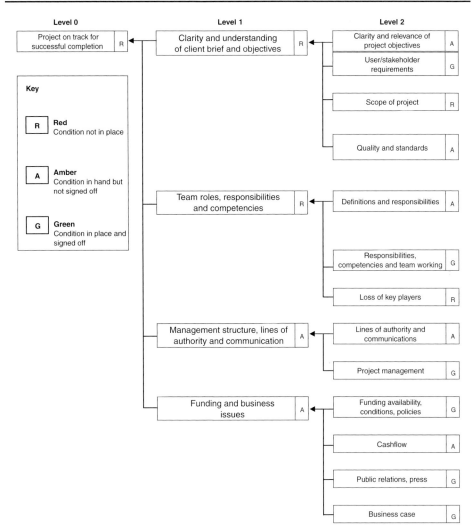

**Figure 11.5** Extract from Risk Breakdown Structure for common causes of project failure, showing traffic light system of status identification

standard risk register but displays their dependencies in a graphical manner (not dissimilar to a work breakdown structure).

If one applies the commonly used 'traffic light' system to describe the status of each risk (Green = acceptable; Amber = under active management to reduce it to acceptable levels; Red = unacceptable/ unmanaged) the RBS can provide a powerful tool for reporting the status of a project, from a risk standpoint, in regular project reports.

Figure 11.5 illustrates the method using the categories for project failure referred to in Section 6.3.

## 11.9 Failure modes and effects analysis

Chapter 5 referred to failure modes and effects analysis (FMEA) as a technique that was developed in another industry that, with little adaptation, can be used in connection with buildings.

Failure modes and effects analysis is a methodical process for identifying potential causes of failure, assessing the likelihood of their occurrence, their severity and the ease with which they may be detected. This analysis then enables the design team to address and reduce the risk of failure. The relevance of this technique in building is twofold.

First, it may be used in exactly the same way as it is in manufacturing, to minimise the likelihood of failure in the building's systems. This could be of particular relevance for organisations that rely heavily on their systems to conduct business. An example would be a financial trading floor where the consequence of failure could be very costly.

Second, it may be used to identify and assess more general risks facing the occupants of a building.

The method fits well with function analysis, since it explores the reliability with which components perform their functions. The method is used because the presence of a large safety factor does not automatically lead to high reliability. Rather it can lead to an unreliable but overdesigned, and therefore expensive, component. By focusing on how a component may fail to deliver the functions expected of it, the user or designer can redesign it to minimise the likelihood and impact of failures.

There are five main types of FMEA, all of which use the same techniques. These are *system*, concerned with the overall system; *design*, concerned with the functions of a system or subsystem; *process*, concerned with the manufacture and assembly (or building) processes; *service*, concerned with service functions; and *software*, concerned with software functions.

The method uses a table, not unlike a risk register, to record the following information:

■ Description of the component and the function to be performed.

■ How it might fail (failure modes).

■ What would happen if it failed (the potential consequence of failure).

■ The severity of the impact of that consequence (usually on an empirical scale of 1 (low) to 10 (high)).

■ Potential causes or mechanisms of failure (there might be several for each mode).

■ The likelihood of each of the causes occurring (again on an empirical scale of 1 (low) to 10 (high)).

■ The current controls in the design to prevent the causes of failure.

■ The ease with which the cause of failure can be detected (on an empirical scale of 1 (easy) to 10 (difficult)) .

■ The resulting risk priority number (RPN) which is the product of the impact, likelihood and ease of detection.

The RPN gives a priority rating for identifying improvement action; the higher the RPN, the higher being the need for action.

The team can then agree what actions may be taken to address those failure modes with a high RPN, allocate the action to an action owner and agree a timescale by which the action should be completed.

Regular reviews of the FMEA throughout the design process will result in a more reliable and robust design without unnecessary overdesign.

## 11.10  Hazard and operational studies (HAZOP)

This technique is used by operators of a facility to ensure that it is safe to use. The method is a systematic exploration of the components within a system using a series of guide words to trigger identification of potential hazards and operability shortcomings.

The first step is to break the system down into components or subsystems (here, function analysis can help). The study leader then poses a series of guide words to each component and records any deviations from ideal behaviours that may be identified by the team. He will also note the cause of the deviation and its consequence (or severity – see below). Finally, the team will identify any actions that may be needed to reduce unacceptable levels of hazard or operability (see Figure 11.6).

**Consequence category**

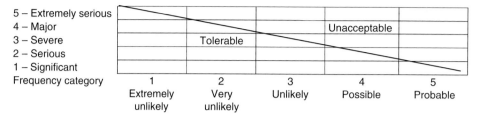

Figure 11.6 HAZOP action assessment table

Commonly used guide words are: more, no, less, reverse, other and also. Actions are assessed from a table similar to a qualitative risk analysis table by placing the deviation into one of the five categories and assessing the likely frequency of occurrence. An example of a HAZOP analysis form is included in Chapter 12.

# 12 Toolbox – checklists, forms and tables

## 12.1 Toolbox – summary

When delivering value and risk management studies, the study leader's tasks will be made a lot easier if he has to hand array of proformas, checklists and other such tools. This chapter sets out a number of the more commonly used tools with a short explanation of when he can use them. None of them is prescriptive. He can use them to develop a format that suits his style.

## 12.2 Briefing checklist

Every study starts with a briefing with the key stakeholders to enable the study leader to understand the project and set up the study appropriately. This checklist provides some of the heading she or he should address.

- ■ *To understand the project*
    - o The project objectives
    - o The expected benefits, measures and targets
    - o Scope and description
    - o Stakeholders and management set up
    - o Critical issues
    - o Timescale
    - o Plans and drawings
    - o Cost summaries
    - o Stage in the project lifecycle, timing and phasing

- o   Value drivers

- o   Basis for whole-life costing.

- ■   *To understand the type of study that is needed*

  - o   Objectives of the study

  - o   Scope and boundaries (givens)

  - o   Success criteria and targets

  - o   Issues, concerns and constraints

  - o   Risks

  - o   Information needed, availability and where it may be found

  - o   Analysis and reconciliation of information

  - o   Wrap-up meeting for presentations to decision-makers

  - o   Report who is it for, timing, presentation and numbers.

- ■   *Study logistics*

  - o   The study team

  - o   Timetable and structure

  - o   Venue and facilities

  - o   Communication

  - o   Distribution of information (including workshop invitation and risk identification forms).

- ■   *Follow-up*

  - o   Who will track recommendations and how

  - o   How to measure the achievement of targets

  - o   Who will undertake project reviews and when.

## 12.3  Information checklist

An effective study requires that certain key information is available to the study leader, for analysis, and the study team for information. The following checklist provides a basis for gathering it:

- ■   The project brief

- ■   Statement of requirements

- Background and history of development

- Performance data

- Specifications

- Drawings, models

- Programmes and key stages

- Current status of project

- Process flow diagrams

- Cost information, capital, operating and lifecycle

- Revenue information, disposal salvage

- Previews reviews/reports, feedback from similar jobs

- Feasibility studies, option studies

- Planning and other statutory approvals

- Project and other meeting highlights

- Project directory and communication plan

- Other information identified in briefing meetings.

## 12.4 Study toolkit

Before undertaking a study the leader should check that he has the appropriate visual aids and equipment. The following list provides a good starting point:

- Laptop and power lead, mouse, printer drivers

- Data projector, connection leads

- Software, custom files, disks/CDs

- V-RM general and study specific presentation files

- Feedback from pre-study consultations and analysis

- Printer, printer lead

- Paper, transparencies, flipcharts

- Pens for paper, whiteboards, flipcharts, transparencies, high-lighters

- Self adhesive notes, adhesive putty, drawing pins, tape, stapler, etc.

- Workshop handbook/invitation and study information

- A0 enlargements of function analysis or other input material

- Attendance forms and feedback forms

- Proposal development forms

- Sample report

- Sounder to call the meeting to order (optional).

## 12.5 Workshop invitation

The study leader should include the following information in any invitation to a workshop to ensure that all participants come prepared:

- Name of the project

- The purpose of the workshop

- Overview of the process to be followed

- The scope of the workshop

- The agenda

- What preparation the participants should do

- What information they should have.

## 12.6 Recording information generated during the workshop

It saves considerable time and provides consistency of output if the study leader prepares forms (in soft and hard copy) for gathering the information generated during the workshop.

### *Value management idea listing and evaluation form*

Typically this would contain the following headings and columns:

- The study to which it relates

- Idea evaluation criteria and ranking method

- Unique identification number of the idea including category

- Outline of the idea

- Evaluation ranking
- Summary of estimated impacts (savings, extras, other)
- Cross-references to other similar ideas
- Comments.

## *Value management proposal development summary form*

- Details of the study to which the proposal relates
- Number of idea(s) to which the proposal relates
- Date of preparation and owner
- Outline description of idea to which it relates
- Description of the existing solution and functions performed
- Description of the proposed alternative and functions performed
- Advantages of adopting the new proposal
- Disadvantages of adopting the new proposal
- How it would be implemented
- Impact on whole-life cost, time and quality (functionality)
- Discussion at presentation meeting
- Recommendations for implementation.

The proposal summary development form may be supported by back-up sheets for showing sketches, calculations and the like.

## *Risk identification, assessment and management form*

Typically this would contain the following headings and columns:

- The study to which it relates
- The status of the form (identification, assessment, management, review, etc.)
- Brief description of the risk
- Unique number
- Category

- Consequence arising should the risk occur

- Likelihood of occurrence

- Impact (on cost time or quality)

- Qualitative rating

- Columns for recording probability and impact estimates if risks are to be quantified

- Action owner

- Management action

- Date by when effectives should be reviewed.

## *Other proformas*

Other useful proformas include:

- Form for listing issues and observations made during the workshop.

- Form for listing actions arising from the workshop.

- Study feedback forms to assess what went well, what could be improved and so on. This helps the study leader to improve the services in future studies.

- Attendance lists, showing who attended on what days.

## 12.7 Reports

A key deliverable form of all studies is a formal report. This provides a record of the outcomes as well as the basis for future decisions and management actions. A report should contain the following minimum information. The report may be 'layered' to enable all or parts of it to be distributed to different audiences (e.g. summary for senior management, summary plus study details for middle management, summary plus study details and appendices containing calculation sheets for the project delivery team).

- The study title and the project to which it relates

- Clear indication of the context of the study in terms of any overall programme

■   The purpose of the study

■   A summary of the outcomes

■   A project synopsis (to allow a third party to understand the context of the study)

■   The study contributors, including any external stakeholders consulted

■   A summary of the briefing meetings and consultations programmes

■   The information on which the study was based

■   Feedback and analysis of information gathered during consultations

■   The study process

■   Details of information received or presented

■   Details of information generated during the study and any workshops

■   Detailed deliverables, including value management proposals, risk registers

■   Implementation plan details including who is responsible for the programme, action owners, actions, review dates, criteria for judging success, escalation, etc.

## 12.8  Value management categories/prompt list

The following categories can be used as a basis for seeking ideas for value improvement. This could be expanded into a detailed checklist. This risks stifling original and innovative thought.

*Business plan.* Understanding client's business plan to increase revenue, responsiveness or efficiency.

o   Increase output/performance/revenue/aesthetics/reputation

o   Relax constraints, challenge spec/brief, client decisions

o   Simplify/improve (business) process

o   Improve/review efficiency/availability

o   Establish value/financial criteria

o   Establish/reconcile with demand/market/need

o   Change of location.

*Certainty.* Increasing certainty.

o   Reduce risk

o   Do trials, test, validate and research

o   Clarify definitions

o   Improve communication

o   Educate users/public

o   Identify actions to enable progress

o   Review method of procurement.

*Capital cost*

o   Eliminate/reduce unnecessary/under-used elements

o   Add/omit

o   Reduce cost

o   Vary quality/material/size to suit differing needs

o   Same element performing more than one function

o   Reduce/change scope

o   Re-use existing component.

*Whole-life cost.* Optimising whole-life costs (capital and operating)

o   Reduce whole-life cost (could add capital cost)

o   Extend life

o   Automate

o   Size for today/ignore peaks

o   Allow for expansion

o   Change orientation/layout.

*Time.* Optimising time to completion/timing of completion

o   Review programme, longer, shorter, phasing.

*Excellence.* Supporting design excellence

o   Make more attractive to customers, add delight

o   Create necessary project image/branding.

*Design.* Design/layout considerations

o   Enhance environment for users

o   Innovate/new technology/different technology

o   Health and safety

o   Functional/design efficiency/operability.

*Flexibility.* Making change easier

o   Adaptability

o   Enabling future expansion

o   Exit strategies.

*Buildability.* Ease of construction

o   Simplify element

o   Simplify construction method

o   Standardise/less bespoke.

*Tax.* Property taxation considerations

o   Capital tax allowances for machinery/plant

o   VAT.

*Environmental.* Sustainability and environmental issues

o   Enhance environmental impact/BREEAM/green issues

o   Enhance sustainability (environmental)

o   Enhance social impact

o   Enhance economic impact – globally.

## 12.9  Risk identification prompt list

The following prompt list should be used towards the end of the risk identification process. Earlier use tends to stifle original thought and can lead to long lists of trivial risks that pose no real threat to the project under study.

1. Funding and business issues

   o Budget

   o Funding

   o Affordability

   o Business case

   o Tax.

2. Clarity and understanding of client brief and objectives

   o Clarity of project objectives

   o User/stakeholder requirements

   o Briefing documentation

   o Client standards.

3. Team roles, responsibilities and competencies

   o Structure of the client body

   o Decision-making

   o Clarity of team responsibilities

   o Project team resources

   o Loss of key players.

4. Management structure, lines of authority and communication

   o Lines of authority and communications plan

   o Project management, execution plans

   o Change controls

   o Contract strategy.

5. Programme, information release, decision-making, timing and adequacy

   o Logic linked

   o Procurement programme

   o Design programme

    o   Construction and commissioning programme

    o   Schedules of information required.

6.    Third party and external disruptions to operations

    o   Stakeholder analysis

    o   Site security

    o   Pressure groups

    o   Impact of construction works.

7.    Planning and statutory approvals and health and safety

    o   Health and safety regulations

    o   EU directives

    o   Client department regulations

    o   Fire regulations

    o   Building regulations

    o   Public entertainments licence

    o   Requirements of HMRI and other specialist and authorities

    o   Hazardous substances

    o   Tree preservation orders

    o   Listed building consent

    o   Conservation area consent

    o   Scheduled monument consent

    o   SSSI

    o   Rights of way and way leaves

    o   Potential changes in legislation.

8.    Construction, site conditions, ground, weather and access

    o   Existing building

    o   Access

    o   Ground conditions

    o   Temporary works

    o   Security

    o   Utilities, statutory undertakings.

9.   Procurement uncertainties, cost, time or quality

    o   Risk allocation

    o   Suitability of procurement route

    o   Tender market

    o   Tender documentation

    o   Tender prequalification

    o   Cost estimate adequacy

    o   Tender evaluation

    o   Specifications.

10.   Unresolved design issues

    o   Design coordination

    o   Design responsibility

    o   Contractor design adequacy

    o   New technology

    o   Untried design solution

    o   Obsolescence.

11.   Contractor competency and general site management

    o   Site management

    o   Work package interfaces

    o   Method statements

    o   Specialist skills

    o   Realistic programmes

    o   Commissioning procedures and documentation.

12.   Force majeure, natural or manmade disasters

    o   Specific risks to project.

## 12.10 Value profiling proforma

The form shown in Figure 12.1 can be used to establish an organisation's value profile. This can be used as the basis for

| | | | Unacceptable | | | | | | | | | Delight | | |
|---|---|---|---|---|---|---|---|---|---|---|---|---|---|---|
| Value drivers | Measure | Weight | 1 | 2 | 3 | 4 | 5 | 6 | 7 | 8 | 9 | 10 | | Score |
| | | | | | | | | | | | | | | |
| | | | | | | | | | | | | | | |
| | | | | | | | | | | | | | | |
| | | | | | | | | | | | | | | |
| | | | | | | | | | | | | | | |
| | | | | | | | | | | | | | | |
| | | | | | | | | | | | | | | |
| | | | | | | | | | | | | | | |
| | | | | | | | | | | | | | | |
| | | | | | | | | | | | | | | |
| | | | | | | | | | | | | | | |
| | | | | | | | | | | | | | | |
| | | | | | | | | | | | | | | |
| | | | | | | | | | | | | | | |

**Figure 12.1** Value profiling proforma

value-based decision-making, value benchmarking against other similar projects, identifying areas with most potential for adding value or assessing value added through the use of value management.

## 12.11 Option evaluation matrix

The form shown in Figure 12.2 is used to evaluate the relative benefits of a small number of options against selection criteria. Value drivers may be used as the selection criteria.

## 12.12 Weighting tool

The tool shown in Table 12.1 can be used for the simple weighting procedure of distributing 100 points across each of the items to be weighted.

## 12.13 Paired comparison proforma

The proforma indicated in Figure 12.3 may be used when applying the paired comparison method of weighting a selection of items.

| | | | EVALUATION CRITERIA | | | | | | | | | |
|---|---|---|---|---|---|---|---|---|---|---|---|---|
| | | | | | | | | | | | | |
| | | No. | 1 | 2 | 3 | 4 | 5 | 6 | 7 | 8 | | Value |
| No. | OPTION | Weight | | | | | | | | | Total | index |
| 1 | | | | | | | | | | | | |
| 2 | | | | | | | | | | | | |
| 3 | | | | | | | | | | | | |
| 4 | | | | | | | | | | | | |
| 5 | | | | | | | | | | | | |
| 6 | | | | | | | | | | | | |
| 7 | | | | | | | | | | | | |
| 8 | | | | | | | | | | | | |
| 9 | | | | | | | | | | | | |
| 10 | | | | | | | | | | | | |
| | | Cost^ | | | | | | | | | | |
| | Benefit ranking: Poor – 1; Fair – 2; Good – 3; Excellent – 4. | | | | | | | | | | | |

**Figure 12.2** Option evaluation matrix

**Table 12.1** Weighting tool

| Value drivers or criteria | Individual scores | | | | | | | Totals | % weight | Rank |
|---|---|---|---|---|---|---|---|---|---|---|
| A | 40 | 35 | 50 | 20 | 40 | 25 | 25 | 235 | 34 | 1 |
| B | 10 | 5 | 5 | 10 | 5 | 15 | 5 | 55 | 8 | 4 |
| C | 5 | 10 | 0 | 10 | 10 | 10 | 10 | 55 | 8 | 4 |
| D | 10 | 5 | 5 | 10 | 5 | 10 | 10 | 55 | 8 | 4 |
| E | 20 | 15 | 30 | 30 | 15 | 15 | 20 | 145 | 21 | 2 |
| F | 5 | 10 | 5 | 10 | 10 | 10 | 10 | 60 | 9 | 3 |
| G | 5 | 10 | 0 | 5 | 10 | 10 | 10 | 50 | 7 | 7 |
| H | 5 | 10 | 5 | 5 | 5 | 5 | 10 | 45 | 6 | 8 |

## 12.14 Proforma for whole-life cost estimating

Table 12.2 indicates the principal categories of information that are needed to compile and estimate of whole-life costs for a 25 year

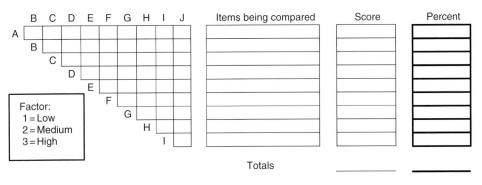

**Figure 12.3** Paired comparison proforma

**Table 12.2** Proforma for whole-life cost estimating

| Costs | Amount £ | Period Years | Year of activity | 1 | 2 | 3 | 4 | 5 | 6 | 7 | 8 | ..... | 24 | 25 | Total |
|---|---|---|---|---|---|---|---|---|---|---|---|---|---|---|---|
| Initial | | | Design and build period | | | | | | | | | | | | |
| Periodic | | | Varies | | | | | | | | | | | | |
| annual | | | Every year | | | | | | | | | | | | |
| (revenue) | | | Every year | | | | | | | | | | | | |
| cashflow | | | Sum of above | | | | | | | | | | | | |
| Discount factor | | | Calculated by spreadsheet | | | | | | | | | | | | |
| Discounted cashflow | | | Calculated by spreadsheet | | | | | | | | | | | | |
| Net Present Value (sum of all discounted cashflows over whole life of asset) | | | | | | | | | | | | | | | |

asset life. The additional information needed to perform a discounted cashflow analysis is the discount rates to be used.

## 12.15 Scenario building form

When building scenarios as described in Chapter 10, a form similar to that indicated in Table 12.3 should be used.

## 12.16 Whole-life cost checklist

When gathering information for compiling a whole-life cost estimate, the checklist shown in Table 12.4 provides a useful summary of headings under which to gather information.

## 12.17 HAZOP analysis sheet

Figure 12.4 gives an example of a HAZOP analysis summary sheet.

**Table 12.3** Scenario building form

| No | Description | Saving £000s | Extra £000s | Area gain sq. ft. | Value of area gain (NPV) £000s | Comment | Capital cost | Exciting image | Maximise NLA | Optimise occupation costs | Create comfortable environment | Accommodate third-party requirements | Provide basic spaces | Base | A | B | C |
|---|---|---|---|---|---|---|---|---|---|---|---|---|---|---|---|---|---|
| 1 | Reduce no scenic lifts from 5 to 4 | 300 | | 1450 | 800 | Need to reduce waiting times | + | 0 | + | + | – | 0 | 0 | | ✓ | ✓ | |
| 3 | Additional lift | | 200 | –1450 | –800 | Reduces waiting times | – | 0 | – | – | + | 0 | 0 | | | | ✓ |
| 7 | Fix width of glass bridges | 175 | | | | | + | 0 | 0 | 0 | 0 | 0 | 0 | | ✓ | ✓ | |
| 9 | Solid atrium bridges, varying width | 150 | | | | | + | – | 0 | 0 | 0 | 0 | 0 | | | | ✓ |
| 12 | Reduce cladding thickness on west façade | 60 | | 450 | 250 | | + | 0 | + | – | 0 | 0 | 0 | | ✓ | ✓ | |
| 14 | Reduce cladding thickness throughout | 200 | | 1600 | 880 | | + | 0 | + | – | 0 | 0 | 0 | | | | ✓ |
| 43 | Improved cleaning cradle | | 165 | | | | – | 0 | 0 | + | 0 | + | 0 | | ✓ | ✓ | ✓ |
| 52 | Additional louvers at roof level | | 115 | | | | – | + | 0 | 0 | 0 | + | 0 | | ✓ | ✓ | ✓ |
| 92 | Tinted glazing | | 90 | | | | – | + | 0 | + | + | 0 | 0 | | ✓ | ✓ | ✓ |
| 95 | Brize soleil | | 150 | | | | – | 0 | 0 | + | + | 0 | 0 | | | | ✓ |
| 6 | Replace water wall with glass screen | 50 | | | | Reduces running costs | + | – | 0 | + | + | 0 | 0 | | ✓ | ✓ | ✓ |
| 15 | Reduced beams due to additional columns | 250 | | | | | + | 0 | 0 | 0 | – | 0 | 0 | | ✓ | | ✓ |
| 17 | Increased floor spans | | 100 | | | | – | 0 | 0 | 0 | + | 0 | 0 | | | ✓ | |
| 76 | Relax BREEAM from excellent to good | 0 | 0 | | | | 0 | 0 | 0 | – | – | 0 | 0 | | ✓ | ✓ | ✓ |
| 65 | Omit disabled toilets on alternate floors | 15 | | | | | + | 0 | 0 | 0 | – | 0 | 0 | | ✓ | ✓ | ✓ |
| | | | | | | | | | | | | | Value index | 580 | 730 | 720 | 780 |

| Value driver | Wt. | Base case Scale | Base case Score | Proposal A Scale | Proposal A Score | Proposal B Scale | Proposal B Score | Proposal C Scale | Proposal C Score |
|---|---|---|---|---|---|---|---|---|---|
| Capital cost | 20 | 4 | 80 | 6 | 120 | 5 | 100 | 7 | 140 |
| Exciting image | 20 | 8 | 160 | 9 | 180 | 9 | 180 | 9 | 180 |
| Maximise NLA | 30 | 5 | 150 | 7 | 210 | 7 | 210 | 8 | 240 |
| Optimise occupation costs | 10 | 6 | 60 | 7 | 70 | 7 | 70 | 7 | 70 |
| Create comfortable environment | 10 | 7 | 70 | 8 | 80 | 9 | 90 | 8 | 80 |
| Accommodate third part | 5 | 4 | 20 | 6 | 30 | 6 | 30 | 6 | 30 |
| Provide basic space | 5 | 8 | 40 | 8 | 40 | 8 | 40 | 8 | 40 |
| Total | | | 580 | | 730 | | 720 | | 780 |

Note: Scale × Wt. = Score

## 12.18 Room layouts

The layout of the room can make a big difference to the atmosphere and interactions that take place within the workshop. Two popular layouts are shown in Figures 12.5 and 12.6. The first shows a layout that is suitable for informal workshops or events such as

**Table 12.4** Whole-life cost checklist

| | Category |
|---|---|
| **1.0** | Initial and capital costs |
| | Site acquisition costs |
| | Site preparation costs |
| | Construction costs |
| | Statutory undertakers |
| | Fit out costs |
| | Fixtures, fittings and equipment |
| | Communications and IT |
| | Professional fees |
| | Client management resources |
| | Moving and relocation costs |
| | Taxation |
| | General overheads |
| | Energy |
| | Utilities |
| | Consumables |
| | Staff or contract staff |
| | Security |
| **2.0** | Annual costs |
| | Routine planned maintenance and replacement of elements through wear and tear |
| | Relocation costs |
| | Finance costs |
| | Tax |
| **3.0** | Periodic costs |
| | Alterations |
| | Replacements |
| | Capital allowances relating to alterations or improvements |
| **4.0** | Income |
| | Income |
| | Residual values |
| | Savings |
| | Disposal |
| | Tax |

stakeholder conferencing. Another favourite is a U-shaped layout with the participants placed around the outside of the U shape, each with a good view of the facilitator at the apex of the U. This layout enables the facilitator to move about the room and speak to individuals face to face as the workshop progresses. It is, however, less appropriate for breakout activities, whereas the circular tables indicated upon can enable the team to break into groups without changing rooms.

| No. | Element | Secondary function | Guide word | Deviation | Possible causes | Consequences | Safeguards | Comments/ likelihood | Actions required | Action by | Date by |
|-----|---------|--------------------|-----------|-----------|-----------------|--------------|------------|----------------------|------------------|-----------|---------|
|     |         |                    |           |           |                 |              |            |                      |                  |           |         |
|     |         |                    |           |           |                 |              |            |                      |                  |           |         |
|     |         |                    |           |           |                 |              |            |                      |                  |           |         |
|     |         |                    |           |           |                 |              |            |                      |                  |           |         |

Study title:                                   Sheet:

Reference drawing no.:        Revision no.:    Date:

Team composition:                              Meeting date:

Part considered:

Primary function:

**Figure 12.4** HAZOP analysis sheet

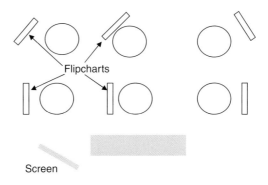

**Figure 12.5** Informal workshop layout

**Figure 12.6** U-shaped formal workshop layout

# Appendix A:
# Bibliography

■ Review of Procurement and Contractual Arrangements in the UK Construction Industry, Final Report

■ APM Risk Management Specific Interest Group (2004), Project Risk Analysis and Management (PRAM), Association for Project Management

■ B.S. Technical Committee MS/2 (2000), BS6079 : 2000 Project Management – Part 3 – Guide to the Management of Business Related Risk, B.S.I.

■ Barton, R (2000), Soft Value Management Methodology for Use in Project Initiation – A Learning Journey, Journal of Construction Research

■ Barton, R (1995), Achieving Participation in Large Groups, University of Canberra

■ Bernstein, Peter L (1998), Against the Gods: The Remarkable Story of Risk, John Wiley & Sons Inc

■ Best, R and de Valence, G (1997), Building in Value: Pre-design Issues, Arnold

■ CIOB (2002), Code of Practice for Project Management for Construction and Development, 3rd edition, Blackwell Science

■ Constructing Excellence (2004), Construction Industry Key Performance Indications, Constructing Excellence

■ Davis Langdon Consultancy (2003), The Social and Economic Value of Construction, Davis Langdon, LLP

■ Davis Langdon Consultancy (1996), Value Management in Construction, CIRIA

■ Davis Langdon, LLP (2004), Getting Value for Money from Construction Through Design – How Auditors Can Help,

National Audit Office; Commission for Architecture and the Build Environment and Audit Commission; OGC

- Egan, Sir J and Construction Task Force (1998), Department of the Environment, Transport and the Regions, Rethinking Construction Ltd

- Fisher, R and Ury, W (1987), Getting to Yes: Negotiating Agreement Without Giving In, Arrow Books Ltd

- Flanagan, R, Ingram, I, Marsh, L and Reading Construction Forum (1998), A Bridge to the Future: Profitable Construction for Tomorrow's Industry and its Customers, Thomas Telford

- Fowler, TC (1990), Value Analysis in Design, Van Nostrand

- Godfrey, PS and Sir William Halcrow and Partners Ltd (1996), Control of Risk: A Guide to the Systematic Management of Risk from Construction, CIRIA

- Goldratt, EM (1994), Its Not Luck, Gower

- Green, SD (1992), A SMART Methodology for Value Management, The Chartered Institute of Building, CIOB

- Green, SD (1999), Going Against the Grain – Critical Perspectives on Construction Management, Unpublished Seminars

- HM Treasury (2004), The Orange Book: Management of Risk Principles and Concepts, HM Treasury

- HM Treasury (1997), The Green Book – Appraisal and Evaluation in Central Government, T.S.O.

- Hunter, G and Stewart, RB (2002), Beyond the Cost Savings Paradigm – Evaluation and Measurement of Project Performance – Value World, Vol 25, No.1, Spring 2002, Society of American Value Engineers (SAVE)

- Institute of Value Management and BSI Technical Committee DS/I (2000), PD 6663: Guidelines to BS EN 12973: Value Management – Practical Guidance to its Use and Intent, British Standards Institution

- Institution of Civil Engineers, The Faculty and Institute of Actuaries (1998), Risk Analysis and Management for Projects (RAMP), Thomas Telford

- Jackson, S (2004), The Management of the Risk – Yours, Mine and Ours, Royal Institute of Chartered Surveyors

- Kaufman, J (1998), Value Management Competitive Creating Advantage, Crisp Management Library

- Kelly, J, Male, S and Graham, D (2004), Value Management of Construction Projects, Blackwell Science

- Kelly, J, Morledge, R and Wilkinson, S (2002), Best Value in Construction, Blackwell Science

- Latham, Sir M (1994), Constructing the Team: Joint, HMSO

- Loughborough University and Development Partners (2005), Value in Design (VALiD), Loughborough University

- Weatherhead, M, Owen, K and Hall, C (2005), Integrating Value and Risk in Construction, CIRIA

- Morris, PWG (1997), The Management of Projects, Thomas Telford

- Office of Government Commence (2003), Achieving Excellence in Construction Procurement Guide 04 Risk and Value Management, OGC

- Office of Government Commerce (2002), Management of Risk: Guidance for Practitioners, TSO

- Office of Government Commerce (2004), Successful Delivery Tool Kit, OGC website www.ogc.gov.uk

- Smith, NJ (1999), Managing Risk in Construction Projects, Blackwell Science Ltd

- Stevens, D (1997), Strategic Thinking Success Secrets of Big Business Projects, McGraw Hill, Australia

- Thiry, M (1997), Value Management Practice, Gilmour Drummond Publishing

- Trafford, DB (1995), Achieving Sustained Value Through Alignment of Operational Behaviour to a Chosen Value proposition – Value, Vol 4, No. 3, Institute of Value Management

- Walker, P and Greenwood, G (2002), The Construction Companion to Risk and Value Management, RIBA Enterprises

- WS Atkins Consultants and Gilberson, Alan (only 2nd edition) (2004), 2nd edition, CDM Regulations – Practical Guidance for Clients and Clients Agents, CIRIA

# Appendix B: Glossary

This section provides a short description of the meaning of terms that are used within this book. The terms may differ from the meanings of the terms in other contexts.

| | |
|---|---|
| **A Post-occupation project review** | Review undertaken after handover to assess the success of the project or processes within it. |
| **Action owner** | The person responsible for ensuring an action is undertaken |
| **Actuarial data** | Data based upon historical records |
| **Adding value** | Increasing the value ratio (benefits/resources used) in a project |
| **Affordability** | Ability to pay for work as it proceeds (cashflow as well as quantum) |
| **Allocation of risk** | How responsibility for risk is split between the contracted parties |
| **Anchor study** | Study at the outset of a programme of projects to align parameters |
| **Anecdotes** | Short case study |
| **Annual costs** | Costs incurred in a single year |
| **Appetite** | Willingness for a person or organisation to accept something (e.g. risk) |
| **Assessment matrix** | Array to set out the basis of comparing attributes |
| **Balanced score card** | Method of assessing an organisation's performance proposed by Kaplan and Norton |
| **Benchmarking by value** | Comparing projects by value |

| | |
|---|---|
| **Benefit risk** | Risk to achieving benefits in full, arising from a risk occurring in a project |
| **Benefit trading** | One organisation using its market strength to gain benefit for another and itself |
| **BRAG** | Acronym for black red amber green colour coding for risks |
| **Breakdown** | Orderly separation of components into their separate parts |
| **Brief development study** | Study to develop clarity in the brief (usually during the strategy or feasibility stages of a project) |
| **Business and political risk** | Risk to client's business or reputation arising from a risk occurring in a project |
| **Business continuity** | Ability to continue business without interruption |
| **Business objectives** | The aims of doing business |
| **Case studies** | Examples of the application of a process |
| **Cashflow** | The sum of income-less costs over a given period |
| **Cause and effect diagrams** | Method of identifying risk |
| **CDM regulations** | Construction, design and management regulations (to ensure that a facility is safe and healthy to build and use) |
| **Charter** | Accord between parties agreeing to certain principles |
| **Communications plan** | Document setting out how people and organisations will communicate with one another |
| **Consensus** | Form of agreement between individuals or organisations |
| **Consequence** | The result of a risk occurring |
| **Consequential risk** | Risk to a client's operations arising from a risk occurring in a project |
| **Constraint** | Imposed limitation |
| **Contingency** | Allowance set aside or a plan as a precaution against future need |

| | |
|---|---|
| **Continuous improvement** | Year on year improvement |
| **Contractor** | Organisation in contract with another |
| **Cost drivers** | Things or events that cause costs |
| **Critical path** | Project planning term linking activities for which time is critical and there is no slack |
| **Critical success factors** | Things that are essential if something is to succeed |
| **Cycle** | Series of repeated activities |
| **Decision tree** | Diagram linking decisions |
| **Design and cost review** | Type of value management study undertaken late in the pre-construction stage of a project |
| **Design, build finance and operate (DBFO)** | Method of procurement where the contractor funds, designs, builds and operates a facility for a specified period |
| **Design quality indicator** | Method of assessing design quality of a completed facility promoted by the Construction Industry Council |
| **Discounted cashflow** | Method of calculating present day values of future cashflows |
| **Embedding** | Consolidating concepts and skills in an organisation |
| **End users** | The people or organisations who use a facility after handover |
| **Environment** | Can refer to either the natural, physical or business surroundings of a project |
| **Escalation** | Means of referring a matter to more senior management |
| **Esteem value** | A concept of value relating to pride in a facility or object |
| **Estimate** | Quantitative measure of the consequence |
| **Evaluation criteria** | Considerations taken into account when selecting something |
| **Exchange value** | A concept of value relating to the saleability of a facility or object |

| | |
|---|---|
| **Exposure** | The potential result of a risk occurring |
| **Failure modes and effects analysis (FMEA)** | Method of assessing how something might fail |
| **FAST** | Function analysis system technique |
| **Function** | What something does. A function may be physical (turn handle) or intangible (enhance appearance). It is normally expressed by an active verb and a measurable noun |
| **Function analysis** | Method of analysing the functions of the constituent parts of a project or a product |
| **Function analysis system technique** | See FAST |
| **Function performance specification** | Method of specifying something by defining what it must do |
| **Gateway** | Term used to define the passage between one project stage and another |
| **GIGO** | Garbage in, gospel out. Relating to the false sense of accuracy given by computer analysis of unreliable data |
| **Givens** | Pre-conditions imposed by the client on the scope of a study |
| **Group work** | Working together with others |
| **Hard and soft values** | Terms used to define different types of value |
| **Human dynamics** | The way in which people react with one another |
| **Impact** | Qualitative measure of the severity of the consequence. It may be positive (an opportunity) or negative |
| **Implementation** | Term used for carrying out an action or plan |
| **Initial costs** | Cost incurred in the acquisition, planning and construction, and handover of a facility |
| **Integrated value and risk management** | Merging the activities of value and risk management into one seamless process |

| | |
|---|---|
| **Internal rate of return (IRR)** | Measure of the financial return likely to result from implementing a project |
| **Issue** | A risk that is certain to occur |
| **Issues and observations** | Statements that are made, recorded, and should be addressed during a meeting or a workshop |
| **Key Performance Indicators (KPI)** | Metrics to assess performance |
| **Lead procurer** | The client or owner who initiates the procurement of a facility |
| **Likelihood** | Qualitative measure of the chance that the consequence occurs |
| **Matrix** | Array of numbers and/or words |
| **Maturity model** | Method for assessing competence |
| **Mentoring** | Assisting a newly trained individual to perform a skill |
| **Metrics** | Standard of measurement |
| **Mind maps** | Method of recording linked subjects developed by Tony Buzan |
| **Mindset** | Fixed attitude |
| **Monte Carlo simulation** | Method of computer analysis (of risk) |
| **Need verification studies** | Study at the initiation stage of a project to validate the need for a project |
| **Optimism bias** | Method of assessing risk allowance promoted by HM Treasury |
| **Outcome** | The result of implementing a project or a study |
| **Outputs** | The products of a study |
| **Owner** | The person ultimately responsible for an action or risk, or the ultimate beneficiary of a project |
| **Paradigm shift** | Fundamental change in thinking |
| **Partnering** | Process for collaborative working |
| **Periodic costs** | Costs which arise from time to time during the life of an asset |

| | |
|---|---|
| **PFI** | Private finance initiative – method of procurement of public projects (also known as private public partnership – PPP) |
| **Phases** | Term used to describe splitting up an activity into parts that are undertaken at different times |
| **Plenary** | Coming together of several groups into one |
| **Policy** | High-level overall plan to achieve objectives |
| **Post-occupation reviews** | Reviews undertaken after handover |
| **Primary function** | A function that is directly related to the project objectives |
| **Private finance initiative (PFI)** | See PFI |
| **Probability** | Quantitative measure of the chance that the consequence occurs |
| **Process flow diagram (PFD)** | Diagram to show the stages in a process |
| **Procurement** | Way in which something is bought |
| **Procurement routes** | Different ways in which something may be bought |
| **Profiles** | Method of showing the distribution of value or risk in a project |
| **Programme** | Timetable of activities within a project. Series of value and/or risk management studies and activities throughout the life of a project |
| **Programme definition study** | Study undertaken at strategy or feasibility stage to clarify the definition of a programme of projects |
| **Project** | A defined series of activities intended to bring about beneficial change |
| **Project (delivery) team** | The delivery side supply chain |
| **Project definition** | Clear articulation of the objectives and parts of a project |

| | |
|---|---|
| **Project definition study** | Study undertaken at strategy or feasibility stage to clarify the definition of a project |
| **Project development** | Evolution of a project with time |
| **Project failure** | Failure to deliver the expected benefits in full |
| **Project profile model** | Method of prioritising projects |
| **Project profiling tool** | Tool for prioritising projects |
| **Project review** | Review to assess the progress of a project |
| **Project stage** | Part of a project selected to assist its orderly management |
| **Project success** | Achievement of the expected benefits in full |
| **Prompt lists** | List of previously identified components, activities or events to assist in the identification of project specific ones |
| **Public services** | Services to the public delivered by local or central government or its agencies |
| **Qualitative** | Dimensionless measure of an attribute |
| **Quantitative** | Measure of an attribute which has dimensions (e.g. time or cost) |
| **Quantitative risk analysis** | Calculation of cost or time efects of risk |
| **RAMP** | Risk and management for projects method for managing risk promoted by the Institution of Civil Engineers |
| **Rating** | A qualitative measure of the exposure to risk, the product of likelihood $\times$ impact |
| **Recording** | Task of capturing the outputs from a study |
| **Report** | Summary of the outputs of a study |
| **Risk** | An uncertain event or circumstance that, if it occurs, will affect the outcome of a project |
| **Risk action owner** | The individual or organisation responsible for undertaking the actions to manage a risk |

**Risk allowance**    Quantitative allowance set aside or a plan as a precaution against future need, linked to the risk register

**Risk analysis**    The assessment of the severity of a risk

**Risk management**    The process of controlling the impact of risk

**Risk manager**    The person responsible for leading the risk management process

**Risk register**    A database of risks containing a summary of the information used for managing risk

**Risk response**    Action taken to reduce the exposure to a risk

**Scenarios**    Group of compatible proposals

**Seasoned professional**    Very experienced and competent individual

**Secondary, tertiary functions**    Functions which are subordinate to primary functions

**Sensitivity**    Varying parameters used in a calculation to demonstrate robustness of its outcome

**Shopping lists**    Lists of proposals of varying attractiveness from which the project team can choose

**Soft value management**    Value management that focuses on people-related issues

**Stakeholder analysis**    Method of assessing the attitude and influence of people in relation to a project

**Stakeholder conferencing**    Method of engaging interested parties to debate a project

**Stakeholder management**    Method of managing the expectations and attitudes of stakeholders

**Strategic**    Relating to the high-level planning of a project

**Study**    A combination of activities including preparation, analysis workshop(s), decision building, reporting and implementation within the context of value or risk management

| | |
|---|---|
| **Study leader** | A qualified practitioner who organises and facilitates a value management (or risk management) study or programme of studies, or individual responsible for planning and conducting a study |
| **Study types** | Different types of (value and risk management) study. |
| **Supply chain** | All the people or organisations involved in the realisation of a project |
| **Target costing** | Method of setting budgets based upon market prices |
| **Targets** | Financial, time or quality aspirations |
| **Team building** | Getting people to work well together |
| **Time risk** | Risk that have an impact on the time to undertake an activity |
| **Top-down risk estimation** | Using high-level guidelines to assess the effects of risk |
| **Training and certification systems** | Systems to transfer skills to individuals resulting in certificates of competence |
| **Utility value** | A concept of value relating to the use to which something is put |
| **Value** | Assessment of the benefits brought by something in relation to the resources needed to achieve it |
| **Value (for money)** | The optimum balance between the benefits expected of a project and the resources expended in its delivery |
| **Value analysis** | A similar technique to value engineering, applied to an existing building or product, commonly abbreviated to VA. Also, a method of analysing value |
| **Value and risk management** | Activities to improve value and reduce uncertainty |

| | |
|---|---|
| **Value articulation** | Expressing client aspirations in terms of the constituent parts of the expected benefits resulting from a project |
| **Value cascade** | Diagram linking project objectives to the component parts of a project |
| **Value cycle** | Series of value management activities repeated through the project cycle |
| **Value driver** | A functional attribute that is necessary to fully deliver the expected benefits from a project (equivalent to a primary function) |
| **Value engineering** | A specific technique which may be used to improve the value of an existing design, commonly abbreviated to VE. It is a method of analysing and improving value |
| **Value for money** | Ratio of benefits to investment |
| **Value index** | Dimensionless measure of value for a project, not confined to cost and time |
| **Value management** | An umbrella term used to embrace all activities and techniques used in the effort to deliver better value for the client, commonly abbreviated to VM |
| **Value management study** | A process involving gathering and analysis of information (preparation), one or more workshops to process information, a report summarising the outcomes and including an implementation plan, all consistent with best value management practice, designed to add value to a project or part of a project |
| **Value measuring process (VAMP)** | Method for measuring value, not confined to cost and time |
| **Value perspective** | Expression describing customers view on an organisation's values |
| **Value profile** | Diagram showing the distribution of value in a project |
| **Value profiling** | Method articulating a client's value expectations |
| **Value propositions** | Method of indicating an organisation's preferred approach to its customers |

| | |
|---|---|
| **Value ratio** | The ratio of benefits to resources |
| **Value tree** | Diagram linking functions or components of value |
| **Values** | Individual's or organisation's attitudes and beliefs |
| **VM programme** | A series of interrelated value management studies applied across a major project or a programme of projects |
| **Weighting** | Method of prioritising attributes |
| **Whole-life costs** | The sum of all cashflows over the life of an asset converted to present costs |
| **Workshop** | A formal facilitated event, involving multiple stakeholders and disciplines, taking participants through a structured process to a prescribed outcome |
| **Wrap-up** | Term used to describe a meeting at the end of a study at which decisions are made |

# Index